Selection and Cross-breeding in Relation to the Inheritance of Coat-pigments and Coat-patterns in Rats and Guinea-pigs

BY

HANSFORD MacCURDY AND W. E. CASTLE

CONTENTS.

	Page.
Continuous *versus* discontinuous variations as factors in evolution	1
Variations of rats in coat-color and coat-pattern	4
Mendelian inheritance of coat-colors and coat-patterns in rats	7
Albinism recessive in relation to all types of pigmentation	7
Latent pigment characters and coat-patterns	8
Modification of hooded pattern by crossing with Irish	11
Modification of hooded pattern by selection	14
Selection for reduced stripe	14
Selection for stripe of increased size	17
Individual spots of guinea-pigs not unit characters	18
Color-patterns selected	18
Dutch-marked series (Series D)	20
Head-spot series (Series H)	23
Statistical analysis of the data for Series D and H	25
Nose-spot series (Series N)	26
Conclusions	32
Tables	34
Bibliography	50

SELECTION AND CROSS-BREEDING IN RELATION TO THE INHERITANCE OF COAT-PIGMENTS AND COAT-PATTERNS IN RATS AND GUINEA-PIGS.

By Hansford MacCurdy and W. E. Castle.

CONTINUOUS VERSUS DISCONTINUOUS VARIATIONS AS FACTORS IN EVOLUTION.

It is generally agreed that the course of evolution is largely influenced by two factors, variation and heredity; but opinions differ as to what sorts of variation have evolutionary significance and as to the manner of their inheritance.

It has been recognized by several investigators that variations are of two distinct sorts. Bateson has called these two sorts of variation continuous and discontinuous; more recently De Vries has called them fluctuations and mutations, respectively.

By continuous variation (or fluctuation) we understand ordinary individual variation within a species. The individuals differ among themselves in size, color, and other structural features. By examining a considerable number of them we can form an idea of what is the commonest (or *modal*) condition as regards each structural feature; and likewise what is the average (or *mean*) condition.

Usually, but not always, the modal and mean conditions are approximately the same, and any other condition is the less frequent in occurrence, the greater its deviation from them. It follows that the most extreme condition observed is connected with the most usual (or modal) condition by an unbroken series of intermediate conditions, and we may call the series as a whole "continuous." The distribution of the individuals in such a series is governed by the laws of "chance," and may be successfully analyzed by statistical methods.

We commonly think of a "chance" result as something entirely beyond the control of law, but in reality such is not the case. Nothing is beyond the control of law. If a blindfolded person puts his hand into an urn containing a mixture of black and of white balls, it is a matter of chance whether he grasps a black or a white ball; but if he repeats the operation a considerable number of times, it is perfectly certain that he will draw balls of both sorts in approximately the same proportions in which they occur in the jar. The result is a "chance" one, but controlled by a perfectly definite mathematical law.

A "chance result" has been aptly defined as the result of a number of causes acting independently of each other. If this is a valid definition, then a continuous series of variations is due to no single cause but to several mutually independent ones. Some of the causes may be external in origin, others internal; some temporary in their action, others permanent. It should not surprise us, therefore, to find that continuous variations differ greatly in the degree of their inheritance. De Vries, indeed, has maintained that they are not inherited at all, except temporarily; that selection of abmodal variations from a continuous series is unable permanently to modify a race; that the modifications will persist only so long as selection continues, but will speedily disappear when selection is arrested. This conclusion, however, seems to us altogether too sweeping. A *priori* there is no reason to suppose that *all* the causes operative to produce continuous variation are external in origin and temporary in action, as De Vries's conclusion would seem to imply. If there are in operation, in the production of a continuous series of variations, causes internal in origin, resident in the constitution of the germinal substance, so much of the result as is due to those causes should be inherited and so should be permanent. De Vries, we believe, has overlooked this factor entering into the problem. He has assumed that all the causes of continuous variation ("fluctuations") are either external in origin or due to conditions of the germinal substance purely temporary. He holds, we believe rightly, that all inheritance is due to germinal modification; but assumes, we believe without sufficient warrant, that permanent germinal modification is not a factor in the production of fluctuations.

Another category of variations, discontinuous variations (which include the mutations of De Vries), is considered by Bateson and De Vries as the true and only expression of permanent germinal modification. But, granting the truly germinal origin of mutations, it does not follow that they are the *only* product of germinal modification.

A discontinuous variation, as the name suggests, is unconnected by intermediate conditions with the usual (*modal*) condition of the species. It represents a change, more or less abrupt, from the modal condition of the species, and is strongly inherited, a fact which indicates clearly its exclusively germinal origin.

In the category of discontinuous variations belong abrupt changes in pigmentation and hairiness among both animals and plants, changes in the number of digits or of the number of phalanges in a digit among vertebrates, in the presence or absence of horns among animals and spines among plants, and other similar conditions.

Such changes are not the result of selection; they often appear, as it seems, spontaneously, and they are permanent in the race, if isolated.

De Vries maintains that all species-forming variations are of this sort; that selection is unable to form new species, because it can neither call into existence mutations nor permanently modify a race by cumulation of abmodal fluctuations. Darwin, on the other hand, and the great majority of his followers, while admitting that races are occasionally produced by discontinuous or "sport" variation, ascribe evolutionary progress chiefly to the cumulation through long periods of time of slight individual differences, such as De Vries calls fluctuations. The issue between the two views is sharp and clear. According to De Vries, if we rightly understand him, selection is not a factor in the *production* of new species, but only in their *perpetuation*, since it determines merely what species shall survive; according to the Darwinian view, new species arise through the direct agency of selection, which leads to the cumulation of fluctuating variations of a particular sort.

De Vries and the Darwinians differ not only as to the part which selection plays in evolution, but also as to the nature of the material upon which selection acts. According to De Vries, species are not modified by selection; mutations *are* new species and selection determines only what mutations shall survive, fluctuations having no evolutionary significance. On the Darwinian view, all species, whether arising by mutation or not, are subject to modification by selection.

A great deal can be said in favor of each of these contrasted views, but discussion is at present less needed than experimental tests of the views outlined. To De Vries we owe much for showing that such tests are possible.

It was our purpose to make tests of this sort when we undertook the experiments described in this paper. The questions to which principally attention has been directed are these: (1) Can discontinuous variations be modified by selection alone? (2) Can discontinuous variations be modified by cross-breeding? A negative answer to these questions will support the view of De Vries; an affirmative answer will support the Darwinian view, because it will show that through selection new conditions of organic stability can be obtained; that is, new species may be produced.

The material used consisted of certain discontinuous variations in the color-pattern of rats. The general result obtained is this: Various color-patterns, like the several pigments found in the rodent coat, are mutually alternative in heredity. Each group of individuals referred to the same type of color-pattern forms a continuous series fluctuating in accordance with the laws of chance about a common modal condition. The different types in general do not overlap; they form a discontinuous series. Now, these types may be modified in two different ways: (a) By selection of abmodal variates within the same continuous series, and (b) by cross-breeding between different types. There is no evidence that one of these methods has effects less permanent than the other. So far, then, as these experiments go, they support the Darwinian view rather than that of De Vries.

VARIATIONS OF RATS IN COAT-COLOR AND COAT-PATTERN.

Variations in the pigmented coat of rodents are of two principal sorts: (1) Variations in the character of the pigments found in the coat, and (2) variations in the distribution of those pigments. The character of the pigmentation in the wild rodent is nearly always complex. Two or three pigments are associated together in the same hair, but they differ in their regional distribution, so that a grayish or brown "ticked" coat results, inconspicuous against many backgrounds. The coat of the house-mouse (Bateson, :03) and that of the wild guinea-pig (Castle, :05) contain three optically different pigments—yellow, brown, and black. These all coexist in the same hair. In certain fancy varieties of these rodents, a single pigment is present without the others, or the distribution of the pigments is such that only one sort is conspicuous. Animals pigmented thus are known as black, chocolate, and yellow (or red) varieties. If all three pigments are absent from the coat and likewise from the retina of the eye, a condition known as total albinism obtains.

In rats, rabbits, and certain other rodents, black and yellow self-varieties are well known, but no pure chocolate animal has been observed.

Total albinism and the several "self" conditions of pigmentation are all mutually alternative in inheritance.

Variations in pigment distribution on the body result commonly either from entire absence of pigment from certain regions of the body, in which case the coat has white markings, or from the occurrence of different pigments singly in different body-regions, in which case the body bears spots of different colors. Both sorts of variation may occur simultaneously, in which case the body is spotted with pigments of different sorts and with white.

The color-varieties of rats are fewer and simpler than those of mice, rabbits, and guinea-pigs. Aside from albinos, there are only two "self" (*i. e.*, uniformly colored) varieties, namely, gray (or brown, the color of the wild *Mus decumanus*) and black. Gray is a Mendelian dominant in relation to black.

As regards coat-pattern, there occur two conditions of partial albinism, which differ from each other only in degree, but which may be obtained each probably in a pure (homozygous) condition. These two patterns may be called "Irish" and "hooded." Each occurs either with gray or with black pigmentation. The "Irish" of fanciers, as described by Doncaster (:06),

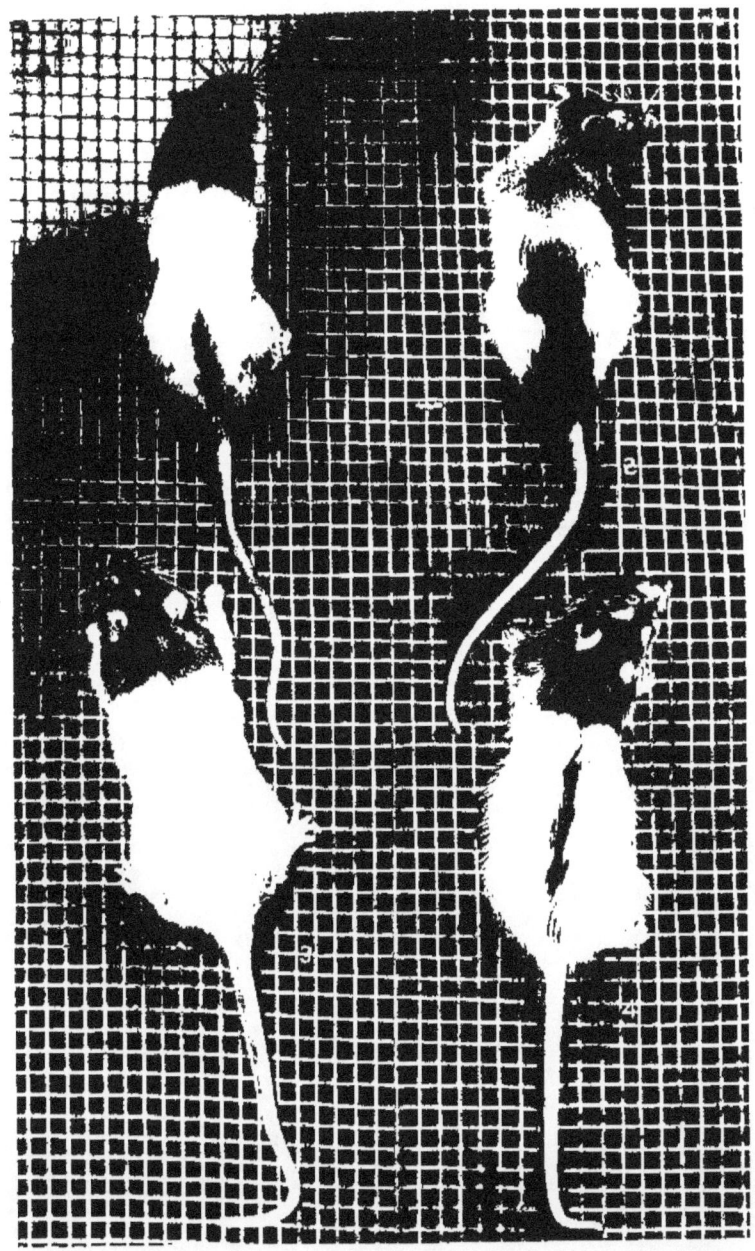

VARIATION IN EXTENT OF PIGMENTATION AMONG CROSS-BRED HOODED RATS.

The usual condition in the original stock is shown in Fig. 4, reduced pigmentation in Figs. 1 and 3, and increased pigmentation in Fig. 2.

pplies only to black-pigmented individuals, but for lack of a better descriptive term, we shall apply it to all animals having the color-pattern of 'Irish" rats, whether gray or black pigmented. A rat of the Irish pattern has pigmented sides and dorsal surface, but bears more or less white fur upon its belly, varying in extent from a few white hairs midway between the front legs to a wholly white ventral surface.

Doncaster (:06) has distinguished two types of Irish rats, in one of which the white is more extensive than in the other. He has observed that the rats with larger white areas are regularly heterozygous, producing hooded as well as Irish offspring when mated *inter se*. Our experiments in the main confirm this idea. It is not possible to determine with certainty, from the size of the ventral patch alone, whether a particular Irish rat does or does not contain the hooded pattern in a recessive condition, but in a lot of Irish rats of similar ancestry those with the larger white patch oftener transmit the hooded condition, while those which do not transmit the hooded condition oftener have a small white patch.

In hooded rats (pl. 1, figs. 1–4) the white areas are more extensive than in Irish ones; pigment occurs only on the head, shoulders, and forelegs (constituting the "hood" of fanciers), and as a median dorsal stripe extending back on to the tail, the stripe being sometimes continuous, sometimes interrupted, and of varying width. According to the nature of the pigment which they bear, hooded rats may be distinguished as gray hooded or as black hooded, precisely as animals bearing the Irish pattern are designated either gray Irish or black Irish.

When crosses are made between rats differing in color-pattern, the more extensively pigmented pattern tends to dominate in the offspring, a fact recognized by Crampe ('77–84, '85), Bateson (:03) and Doncaster (:06). The dominance, however, is not complete, so that the result might be described as "goneoclinic," *i. e.*, intermediate between the parental forms but approximating one much more nearly than the other, in this case always the more heavily pigmented one. Thus, a cross between a wild gray male rat and a black hooded female, known to be homozygous, produced a litter of seven young, all gray, but with a small patch of white on the chest, varying in extent from merely a few white hairs to an area of 4 to 5 sq. cm. The same wild male, mated with an albino female, produced a litter of young similar in character, all with some white below. Bateson (:03, p. 78, footnote) mentions a similar result obtained by Miss Douglas.

Again, a cross between an Irish and a hooded individual produces Irish offspring. An example will be found on page 11 in the matings of black Irish ♀ 141 with black-hooded males, producing nineteen offspring, all Irish. Most matings of Irish with hooded rats have in our experiments produced offspring of both sorts, not because of reversal in the nature of the dominance, but because the majority of the Irish rats used happen to have been heterozygous, bearing the hooded pattern as a recessive character; but, as we have seen, this is not a necessary relationship. Hooded rats, however, bear no other color-pattern, so that any pair of hooded rats of whatever pedigree will produce only hooded pigmented offspring.

To sum up in a brief statement the principal facts presented concerning the coat-patterns of rats, the following series of conditions may be recognized, each with pigmentation less extensive than the preceding: (1) *Self*, whole body pigmented; (2) *Irish*, whole body pigmented except more or less of the ventral surface; (3) *Hooded*, only the head, shoulders, and usually a median dorsal stripe pigmented; (4) *Albino*, no pigmentation.

Each condition behaves as a dominant toward those which follow it in the series, and as a recessive toward those which precede it. Bateson (:03, p. 81) refers to Crampe's failure to obtain the self or the albino conditions by selection from the parti-colored, and adds: "The types are in fact definite, and can not be built up by cumulative selection." The statement applies strictly only to the two extremes of the series, viz, self and albino, but by implication to the others also. Our experiments, however, indicate that it is possible to modify by selection and cross-breeding both the Irish and the hooded conditions, leading to the production of intermediate conditions. We suspect that the same may be true of the self condition also. It has been pointed out elsewhere (Castle and Forbes, :06; Castle, :06), that the theoretical absolute gametic purity, in Mendelian inheritance, probably does not exist in any case. Heterozygosis leads inevitably to modification in character of the conjugating gametes, not simply as regards the entire assemblage of characters, but as regards each character considered separately. The degree of modification is probably indicated roughly by the imperfection of dominance in the heterozygote. Thus, when a perfect blend is obtained, as in the inheritance of ear-length in rabbits, the gametes transmit that intermediate character. But when dominance is very complete, as in a cross between a self and an albino mammal, the segregation of self and albino conditions is very complete among the gametes formed by the cross-breeds. Yet when spotted individuals result from cross-breeding between self and albinos, these spotted individuals form gametes which transmit that same mosaic character. Now, the cross between self and albino rats gives rats of Irish pattern, but quite variable. But our experiments indicate that from variable Irish rats one can by selection obtain rats transmitting no pattern but Irish. The theoretical importance of this is obvious. Cross-breeding and selection combined are means by which we may not only modify existing Mendelian characters, but may even create new ones. They are, then, factors of prime importance in evolution, even in the case of characters which vary discontinuously.

MENDELIAN INHERITANCE OF COAT-COLORS AND COAT-PATTERNS IN RATS.

While pigment character and color-pattern are both inherited in Mendelian fashion, the two are entirely independent of each other. They are separate and uncorrelated unit characters. Accordingly we find that each type of pigmented coat-pattern, namely, self, Irish, and hooded, may occur either with gray or with black pigmentation, the frequencies with which they occur in the respective combinations being governed by the laws of chance and of Mendelian dominance. This will appear in the detailed discussion of the experiments.

Not only may each coat-pattern occur either in a gray or in a black pigmented individual, but it may occur also in an *unpigmented* animal. Paradoxical as this statement may seem, it is capable of abundant proof. The coat-pattern of course is not visible in an unpigmented (albino) rat, but its presence there as a potentiality can be demonstrated as certainly as the occurrence of a recessive character in a heterozygous dominant individual. Nothing but the presence of pigment is necessary to make the color-pattern manifest. This can be supplied by a mating with a pigmented animal.

Specific pigment potentialities (gray, black, or both) are likewise present in albinos. Consequently we must recognize that albinos transmit inactive both pigments and color-patterns; these, however, are unseen and can not be made visible until some lacking substance borne by all pigmented individuals is supplied. Characters transmitted in this inactive state have been termed by one of us (Castle, :05) *latent*, and that terminology will be followed in the present paper.

ALBINISM RECESSIVE IN RELATION TO ALL TYPES OF PIGMENTATION.

That total albinism behaves as a recessive Mendelian character has been recognized independently by a number of investigators, among the earliest being Correns (:01) and Cuénot (:02). The fact has been abundantly verified in the case of mice (see Castle and Allen, :03), rats (Crampe, '77-84; Bateson, :03; Doncaster, :06), rabbits (Woods, :03; Hurst :05; and Castle :05), and guinea-pigs (Castle, :05). The experiments described in this paper corroborate those of Crampe and of Doncaster with reference to rats. The proportions of albinos and of pigmented individuals in mixed litters are close to the Mendelian expectations, indicating neither selective union of gametes nor lessened fertility of certain sorts of unions. Pigmented rats, in which albinism was recessive, when mated *inter se*, have produced 129 albinos to 384 pigmented young, the numbers expected being 128 and 385, respectively. Albinos mated with pigmented individuals, in which albinism was recessive, have produced 201 albinos to 244 pigmented young, or 45.1 per cent albinos, where 50 per cent are expected. The deviation of nearly 5 per cent from expectation in the latter case is probably a matter of chance and would grow less with more extensive observations, for in studying the inheritance of total albinism some observers record an excess of albinos (Castle, :05), others an excess of pigmented individuals (Allen, :04), while others observe very close agreement with expectation (Cuénot, Darbishire). This is as we should expect on the theory of probabilities (see Allen, :04, p. 82).

LATENT PIGMENT CHARACTERS AND COAT-PATTERNS.

Latent transmission of pigment characters through albinos is a matter requiring fuller consideration. Assuming for the time being its correctness, and knowing the well-established facts, (1) that gray pigmentation in rats is dominant over black (Crampe, Bateson, Doncaster) and (2) that every sort of pigmentation is dominant over albinism, we reach the following conclusion. On Mendelian principles we may expect partial albinos to fall into eighteen actually different classes, though visibly the classes number only four, namely, gray Irish, black Irish, gray hooded, and black hooded. These eighteen classes are enumerated in the four central columns of table 1, page 34. Of gray Irish individuals, there should be eight actually different classes as regards gametic output; of black Irish and gray hooded individuals, there should be four classes each; and of black hooded, two classes.

Albinos, though indistinguishable in appearance, should fall into nine different classes as regards the latent pigment characters and color-patterns which they transmit. The nature of these nine classes is indicated in the last column of table 1. The numerals prefixed to the class designations in table 1 indicate the frequencies in which the various classes may be expected to occur as a result of a particular sort of mating, namely, mating *inter se* gray Irish individuals which bear recessive the three characters, total albinism, black pigmentation, and hooded pattern, individuals designated $GI(W.BH)$. The table is based on the assumption that gray Irish individuals of the sort indicated are Mendelian triple heterozygotes, the three independent pairs of allelomorphic characters being pigmentation *versus* albinism, gray *versus* black, and Irish *versus* hooded. In support of the assumption mentioned, it may be said that by proper breeding tests a majority of the hypothetical classes have been shown to exist. Thus, of the eighteen hypothetical pigmented classes enumerated in table 1, all except four have been shown to occur, those four being $GI(W)$, $GI(BH)$,[*] BI,[*] and $BI(W)$[*]. Further, five out of the nine supposed classes of albinos have been proved

---

[*]Demonstrated to exist since the foregoing was written.

occur, those the existence of which has not yet been demonstrated being [BI], W[GI], W[G–IH], and W[GB–I]. It is probable that more extensive
ts would demonstrate the existence of all the missing classes, as no special
arch has been made for them, the demonstrations obtained being for the
ost part incidental to other investigations.

The fundamental assumption on which table 1 rests finds further justification in the numerical proportions in which the various classes are observed
occur, as will appear from an examination of tables 2 and 3 (pp. 35, 36).
these tables are shown the theoretical Mendelian results and the actual
sults of twenty-three different sorts of matings involving the pigments
d color-patterns which have been discussed in the previous pages. Table
includes only matings between pigmented animals; table 3, matings be-
een a pigmented animal and an albino. Throughout both tables, it will
seen, the expected Mendelian results agree quite closely with those actually
served. Two animals possessing the same recessive character have in
case produced offspring bearing the corresponding dominant character.
rther, the proportions of dominants and of recessives observed in mixed
ts of young agree in general quite closely with those expected. The most
riking deviation is an excess of black hooded young (BH) which is seen in
e totals for both table 2 and table 3. This, however, is not large enough
d does not occur with enough uniformity to warrant one in regarding it
other than accidental. The totals for tables 2 and 3 combined are:

	GI	GH	BI	BH	W	Total.
Observed	69	132	130	186	168	685
Expected	56	137.5	136	158	187	...

In order of size the groups are as expected, except the last two, in case of
iich the order is reversed.

In addition to the experiments recorded in tables 2 and 3, the following
servations may be mentioned as corroborative evidence of the Mendelian
havior of pigment characters, color-patterns, and albinism in rats.

Black hooded individuals bearing white recessive, mated *inter se*, give
ro classes of offspring, albinos and black hooded. These albinos have been
own to bear without exception the black hooded pattern in a latent con-
tion. Of the black hooded offspring, there should be two classes—pure
oded (free from albinism) and hooded individuals bearing albinism re-
ssive, the latter being twice as numerous as the former. Out of 22 off-
ring tested, 14 had albinism recessive and 8 were pure hooded individuals.
Matings of pure black hooded individuals with black hooded ones bearing
hite recessive gave all black hooded offspring, 109 in number. One-half
these should be pure hooded, and of the limited number tested this was
und to be true.

The evidence contained in tables 2 and 3 (pp. 35, 36), is presented in the form of summaries only. To give an idea of the sources from which it was obtained, a genealogy may be considered in detail (fig. 1).

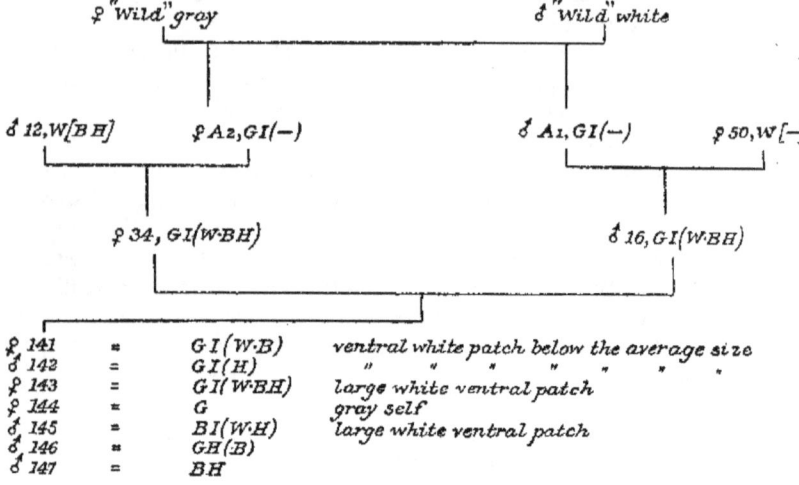

Fig. 1.

A wild gray female and a wild white male were trapped together. Nothing is known of their ancestry, but they were probably derived from escaped tame albinos which had mated with wild gray rats. This pair of captured rats produced gray Irish young, two of which (A_1 and A_2), when mated with ordinary albinos, produced gray Irish young which bore, as recessive characters, total albinism, black pigmentation, and the hooded pattern. Two of these triple heterozygotes (♀34 and ♂16) were now mated together and produced a litter of seven young (141–147, fig. 1), which were tested as to their gametic condition, with the results indicated in fig. 1. Five of the 7 were gray pigmented and 2 black, but no two were alike in gametic character. Males 146 and 147 were mated with black hooded females and with albino females of black hooded parentage, producing nothing but hooded young. This result showed that they bore no coat-pattern except the hooded one, and were free from recessive albinism.

The young of ♂147 were all black pigmented, showing that animal to be entirely homozygous, but some of the young of ♂146 were gray pigmented, some black pigmented, showing that he bore black pigmentation as a recessive character, being otherwise homozygous.

The gray Irish female (141) has produced, by hooded males, young of the Irish pattern only, 19 in number, some of them being gray, and others black pigmented; she has also produced albino young by hooded males bearing recessive albinism. She accordingly bears black pigmentation and total albinism as recessive characters, but is otherwise homozygous, for if she bore the hooded pattern, half her pigmented young should have been hooded. By similar tests, the gametic formulæ of individuals 142 to 145 were established, as given in fig. 1.

Experiments such as these show the complete mutual independence of pigmentation, color-pattern, and total albinism, as has been repeatedly stated. They indicate, further, that partial albino rats (hooded and Irish) regularly form gametes which bear the partial albino condition, either in part or in all of their gametes; and that the extent of the white areas on partial albinos varies in continuous fashion, a self condition being sometimes obtained as an extreme variation of Irish. This last point is important as showing that no complete discontinuity exists between self and Irish patterns in rats, since the gap ordinarily existing between them can be bridged with a complete series of intermediate conditions. The same is true of the Irish and hooded conditions. Though ordinarily discontinuous and behaving as alternative Mendelian characters in inheritance, we can by cross-breeding and selection bridge the gap between them.

MODIFICATION OF HOODED PATTERN BY CROSSING WITH IRISH.

Hooded and Irish individuals differ, as already stated, simply in the extent of their pigmentation. In hooded individuals, the pigmentation is restricted to the hood (which covers the head and shoulders) and to the dorsal stripe; in Irish individuals, the entire coat is pigmented except a ventral patch of variable size. Extension of the dorsal stripe of the hooded pattern, so as to cover the entire dorsal surface and the sides of the body, would yield the Irish pattern. It was our purpose in one set of experiments to see whether such modification could be brought about, and to see also how far an opposite modification of the hooded pattern (further reduction of the pigmented areas) could be effected. Before considering those experiments in detail, we may inquire what effect upon the hooded pattern a cross with the Irish pattern will have.

The hooded stock used in this investigation consisted of individuals which bore albinism recessive, but had no Irish ancestors, so far as known. The extent of the dorsal stripe varied considerably, so that it was found desirable to measure it as accurately as possible in each individual. This was done, first, by estimating the width of the stripe in relation to the total width of the back; next was estimated the extent to which the stripe was interrupted, that is, the proportion of the total stripe length which was pigmented. The product of these two ratios, relative width and relative length of the back-stripe, when multiplied by 100, gives an approximation to the per cent of dorsal surface (posterior to the hood) which was pigmented. Such a valuation, or grade, was obtained for each hooded individual.

The grades for 183 individuals (lot A, fig. 2), belonging to three successive generations, yield the variation curve shown in fig. 3, A. In obtaining this curve grades from 0 to 9 inclusive were grouped in a single class, the mean value of which is 4.5; grades from 10 to 19 inclusive, in a class the mean value of which is 14.5, and so on. The ordinates in fig. 3, A, show the frequencies of these classes. The grades fall into four classes, the one with greatest frequency (modal class) being the one with the lowest mean (4.5), the frequencies of the other classes decreasing in order upward. The average grade for the entire 183 individuals is 13.3 (see vertical line in fig. 3, A).

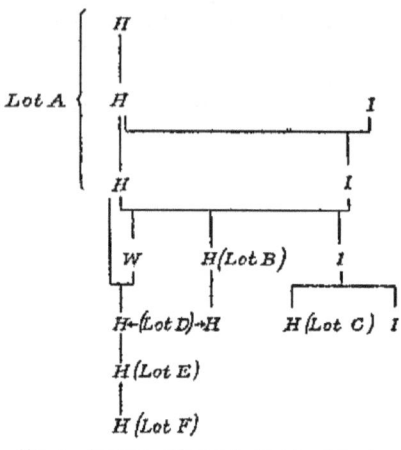

FIG. 2.—Ancestry of lots A to F of hooded rats. Read downward.

The hooded stock was now crossed with an Irish stock, and the Irish cross-breeds thus produced were in turn crossed back with the parental hooded stock. From this last cross were obtained both Irish and hooded offspring, in proportions sensibly equal. These hooded offspring form a group (lot B, fig. 2) derived through one parent directly from the original hooded stock (lot A), but through the other parent from the Irish cross-breeds. One of the two gametes, accordingly, entering into each zygote had a chance to be modified by association with the Irish character. Lot B includes 126 individuals, with back-stripes on the average nearly twice as extensive as those of lot A, the average grades of the two lots being 13.3 for lot A, and 21.2 for lot B. The form of the variation curve for lot B is shown in fig. 3, B. The modal class is higher than in lot A, and the range of variation is extended upward so as to include seven classes, the highest one having a mean grade of 64.5. Accordingly, we conclude that a cross with the Irish stock raises considerably the average size of the dorsal stripe in hooded rats, as well as the range of variation upward in size of stripe.

In the same litters with the hooded rats forming lot B, were born rats of two other sorts, Irish and albino (fig. 2). The former bred *inter se* produced

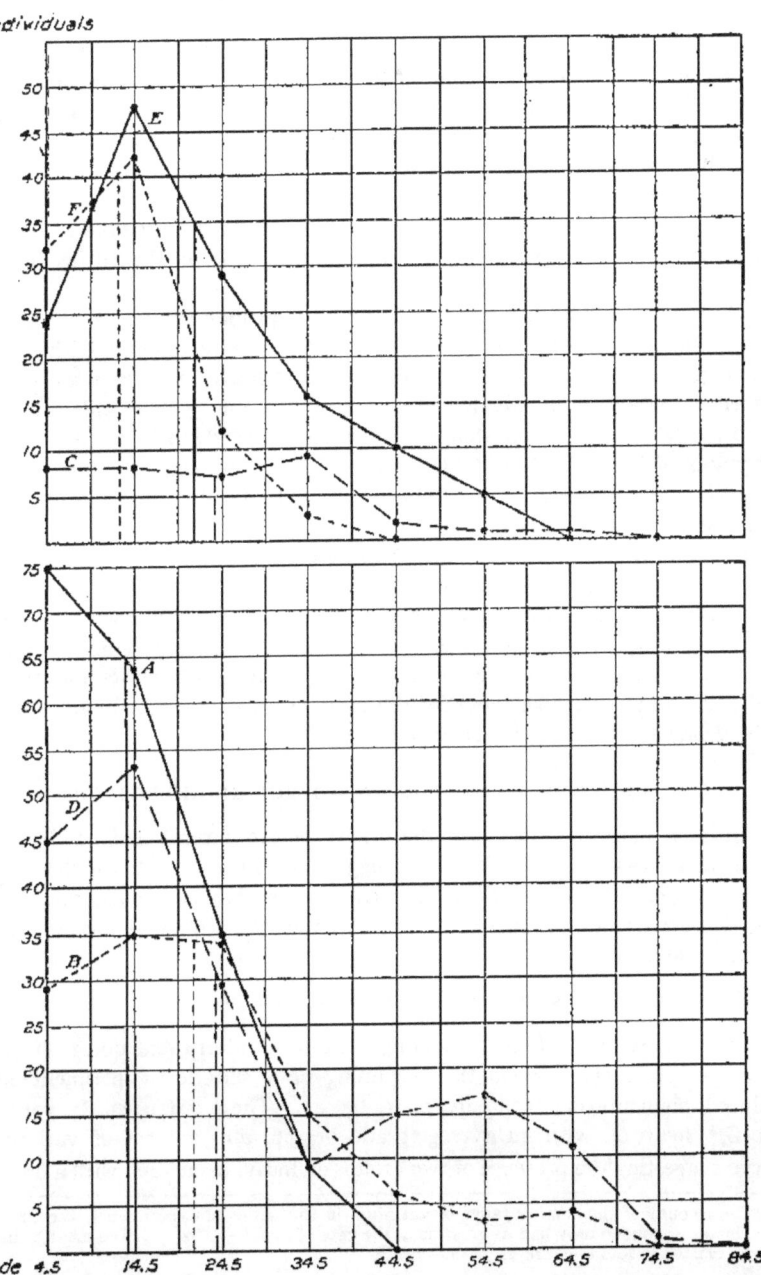

3.—Variation in extent of the back-stripe in lots A to F of hooded rats, showing the effects of crossbreeding with rats of Irish pattern. The positions of the means are shown by vertical lines.

the hooded rats forming lot C (fig. 2); the albinos, when back-crossed with the original hooded stock, produced hooded offspring, which are included in lot D (fig. 2), as are also the hooded descendants of lot B bred *inter se*.

Lot D, accordingly, is a group of individuals of ancestry more mixed than lot B, but on the whole similar. It shows a similar variability (fig. 3, D), but is of a higher average grade and shows a tendency to form a second mode, a group of wide-striped individuals, with a modal grade of 54.5, *i.e.*, individuals in which the stripe is about half as wide as the entire dorsal surface (compare pl. 1, fig. 2).

Lot C (fig. 2 and fig. 3, C) contained only 36 individuals. Its average grade is almost identical with that of lot D, 23.5 as compared with 23.9. Its range also is similar; but the curve itself is flatter, yet with the same tendency to become subdivided into two groups, one with a wider stripe, the other with a stripe like that of the original hooded stock (lot A). Two other lots of hooded rats (E and F, figs. 2 and 3) belonging to this series have also been studied. Lot E consisted of the hooded young of certain individuals of lot D, those having the lesser amount of Irish ancestry, while lot F was derived from certain individuals of lot E bred *inter se*. The parents of lot E had an average grade of 21.8, similar to that of lot B (21.2), or to the lower modal group of lot D; the average grade of lot E itself was very similar, namely 21.1. The parents of lot F had an average grade of 17.8; lot F itself had an average grade of 12.9. The variation curves for lots E and F are unimodal, like those for the original lot A, giving no evidence of a tendency to form a wide-striped variety.

MODIFICATION OF HOODED PATTERN BY SELECTION.

The close agreement between the grades of the parents and children in lot E, and measurably also in lot F, suggests that one can at will either increase or decrease the width of the stripe in a stock of hooded rats. To test this matter more fully, selection has been made both for reduced and for increased size of stripe.

SELECTION FOR REDUCED STRIPE.

Eleven individuals of lot A, having narrow or interrupted dorsal stripes (average grade 11), formed the beginning of a selection experiment for reduced pigmentation (compare pl. 1, fig. 4). They produced 83 young, lot G,* figure 4, with an average grade of 9.6, and a form of variation curve suggesting the presence of two groups of individuals, each with a dorsal

---
*On account of the smaller range of variation in this series of experiments, the size of the classes was made only half as great as in the case of lots A–F, fig. 3. The mean values of the classes in lots G–I are 2, 7, 12, 17, etc.

tripe of different extent, one group being of moderate size (12), the
other extremely reduced (2). Twenty-eight individuals whose grades fell
in the lower part of curve G (fig. 4), and averaged 7.8, were the parents of
the next generation, group H.
This generation consisted of
51 individuals having an average grade of 5.6. The variation curve for this group (H, fig. 4), shows a nearly complete dropping out of individuals with the stripe of moderate size, such as occurred in group G. All the classes above 7 are small and the upward range stops at 22. A few individuals of lot H had no back-stripe at all (compare pl. 1, fig. 3). These with some individuals having a much reduced stripe were selected as parents for the next generation, lot I. The average grade of the parents was in this case 1.7; that of their 34 young (lot I, fig. 4) was 4. Twenty-four of the 34 young fell in the lowest class, mean grade 2, while none were above class 12.

The effect of selection in this series of experiments is clear from a comparison of the four curves, A, G, H, and I, fig. 4. Selection has steadily lowered the average amount of pigmentation in the race by reducing the upward range of the variation curve,

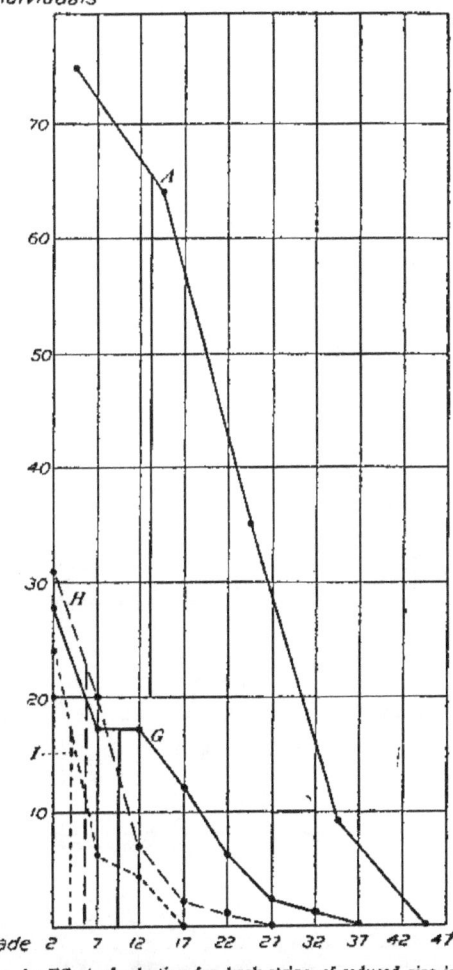

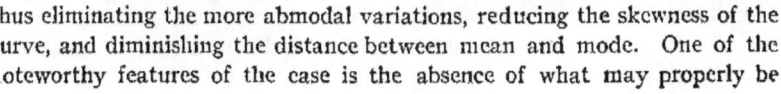

FIG. 4.—Effect of selection for back-stripe of reduced size in hooded rats, lots A, G, H, and I.

thus eliminating the more abmodal variations, reducing the skewness of the curve, and diminishing the distance between mean and mode. One of the noteworthy features of the case is the absence of what may properly be called regression. The filial average does not, with any uniformity, lag behind the parental average, in the process of displacement downward in the variation figure. This fact, together with the decreasing skewness of the variation curve, indicates that the effects of selection in reducing the extent

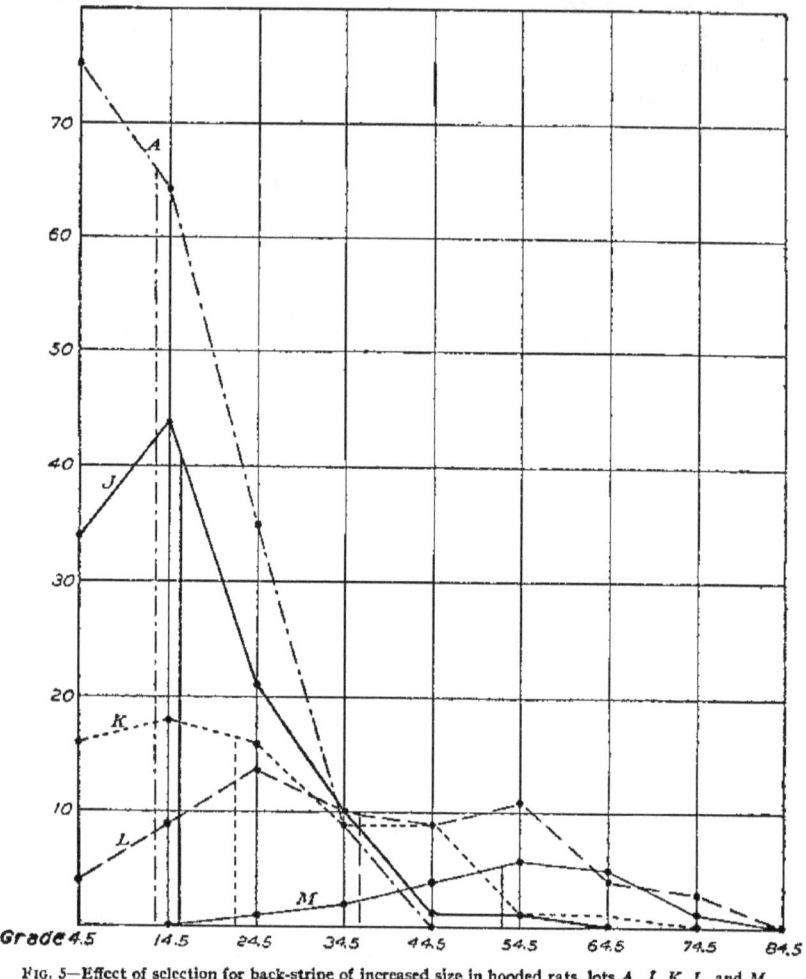

FIG. 5—Effect of selection for back-stripe of increased size in hooded rats, lots A, J, K, L, and M.

of the pigmentation will be permanent, that is, that a stable, narrow-striped variety of hooded rats can be established by selection, and that this variety will breed true.

SELECTION FOR STRIPE OF INCREASED SIZE.

Selection in an opposite direction, for a wider and more continuous stripe, leads to similar conclusions. In this series of experiments, the wide-striped parents used were obtained as a result of crosses between hooded and Irish individuals. Such crosses tended to widen the stripe of the hooded offspring, the widening effect being permanent, as the sequel shows.

Hooded individuals, either parents of wide-striped young, or such as themselves had wide stripes, were selected from lots B and C, and mated together. The average grade of the parents was 12. Their young (lot J, fig. 5), 111 in number, had a higher average grade than the parents, namely, 15.8. The curve was similar in form to the curve (A) produced by the original hooded stock.

A second group of wide-striped individuals, of higher grade than that just described, though of similar ancestry, produced a group of 70 young (lot K, fig. 5), like themselves in width of stripe. The average grade of the parents was 23, that of the young (lot K) was 22.2. The variation curve (K, fig. 5) is a very flat one, similar to curve B, fig. 3.

A third group of wide-striped parents, of still higher grade, was taken chiefly from lot D. Their average grade was 40.7. They produced 64 young (lot L, fig. 5), having an average grade of 36.2. The form of variation curve given by lot L is similar to that given by lot D, flat and bimodal. The upper mode, however, is relatively taller, indicating that the upper (or wide-striped) group is increasing at the expense of the lower, as a result of the selection, and that the original two-peaked condition of curve D was due, not to the heterogeneity of the material included in lot D, but to the fact that *part* of the gametes formed by the cross-bred individuals transmitted a modified (wide-striped) condition. This wide-striped condition was now (in lot L) in process of segregation through the action of selection, and in a fair way to form a stable wide-striped race. Indeed, in the best wide-striped pairs of lot L this seems already to have been accomplished; for certain extremely wide-striped individuals (partly of lot L and partly of a later generation), which had an average grade of 50, produced the 19 young included in lot M, fig. 5, which had an average grade of 53. The mode for this curve is close to the mean, being 54.5, and the curve itself is nearly symmetrical, indicating approach to a condition of stability.

The theoretical importance of the results of these selection experiments is evident. Characters can be permanently modified by selection, contrary to the view of De Vries that such modification is impossible.

The objection may be raised that the permanency of the modification has not been fully established, for later generations might show regression to a narrower-striped condition. While admitting that this objection has some force, the fact may be pointed out that regression does not occur in the experiments made, and if regression were to occur at all, we should expect it to occur and to be greatest in the early stages of a selection experiment. The average grade of the offspring is sometimes greater, sometimes less, than that of their parents, as we should expect if these deviations are purely the result of chance. Table 4 (p. 36) shows the nature and amount of the deviation following each selection made. No great importance attaches to the *exact* amounts of deviation indicated in this table. For in computing the parental averages all parents were weighted alike, but a better method of procedure would have been to weight each parent in proportion to the number of young which it produced. The general nature of the result obtained is, however, not vitiated by the procedure followed; it serves to show the absence of any uniform regression of either a plus or a minus nature.

INDIVIDUAL SPOTS OF GUINEA-PIGS NOT UNIT CHARACTERS.

In an earlier paper (Castle, :05) the fact was pointed out by one of us that in partial-albino guinea-pigs, as in other mammals, the pigmented areas are not distributed without order over the body, but occur in regions fairly definite. These regions in the guinea-pig consist of five paired "pigment centers," dorso-lateral in position. To define their location more accurately, they have been called the *eye, ear, shoulder, side,* and *rump* areas.

The hypothesis was advanced (Castle, :05, p. 45) that these several areas might represent separately heritable unit characters, although experience had clearly shown that, if such were the case, they were not inherited in ordinary Mendelian fashion. To test this hypothesis more fully, we have attempted to establish races of guinea-pigs which should possess certain of the pigmented areas, but should lack others. For this purpose experiments have been made with three different patterns of partial albinism, arbitrarily chosen from those which occur among ordinary spotted guinea-pigs. These are the Dutch-marked pattern (pl. 2, fig. 3), the head-spot pattern (pl. 2, fig. 2) and the nose-spot pattern (pl. 2, fig. 1).

COLOR-PATTERNS SELECTED.

In the Dutch-marked pattern, all the typical spots are present except the shoulder spots. The absence of shoulder spots produces a white belt or girdle round the body. Anterior to the girdle, eye and ear spots are present and usually confluent on the same side of the head, but not across the median plane. Posterior to the girdle, the large side spots and the small rump spots are present and confluent over the back as well as lengthwise of the body, forming a large blanket of pigment covering the whole posterior part of the body, but not extending out upon the feet. This is the ideal "Dutch-marked" type of the fancier. Occasionally it is fully realized, but more often minor departures from the ideal occur, such as asymmetry of the pigmentation of the two sides of the body, failure of the eye and ear spots to coalesce, or entire absence of one of the two, absence of one or both side spots, or the occurrence of a white gap between side and rump spots. Our attempts to fix this pattern by selection, continued through a series of generations, have not been successful.

Animals of the head-spot pattern (pl. 2, fig. 2) differ from Dutch-marked chiefly in the entire absence of pigmentation posterior to the girdle, though the pigmentation on the side of the head also is usually less extensive than in Dutch-marked animals. Because of the practical difficulty of distinguishing eye spots from ear spots, when the two are confluent and similarly pigmented, we have throughout this series of experiments treated the two as forming a single pair, which we call *head spots*.

When the head pigmentation is most reduced, only the retina of the eye is pigmented. Then the head-spot type passes over into the black-eyed white type. In tabulating our results, however, we have treated the two types as distinct. The head-spot pattern, like the Dutch-marked, we have been unable to "fix."

The nose-spot pattern occurs less frequently than the two already described. The entire body is white, except that part of the head which is situated anterior to the eyes. This part of the body is usually unpigmented in spotted guinea-pigs, which fact makes the occasional occurrence of the nose-spot pattern more striking. Our attempts to fix this pattern have succeeded scarcely better than those with the Dutch-marked and head-spot patterns.

In addition to our attempts to fix particular color-patterns arbitrarily chosen, we have made records of the pigment distribution on several hundred spotted guinea-pigs of known ancestry, with a view to ascertaining whether or not particular spots are inherited. The question for which an answer was sought may be stated thus: Does an animal having, let us say, an eye spot, produce young with eye spots any oftener than animals bearing a like amount of pigmentation but without eye spots? If this question receives an affirmative answer, then we may conclude that eye spots, as such, are inherited, and it will remain to ascertain the law governing their inheritance.

The evidence bearing on this question can best be presented in the form of tables, showing what areas of the young are pigmented in comparison with those of their parents. Such a table has been constructed for each of the successive generations in which selection was exercised for the fixation of a particular coat-pattern. By a comparison of the tables for the successive generations, we may learn whether selection was having any effect in fixing the desired coat-pattern. For, if it was doing so, later generations should show closer agreement than earlier ones between parents and children as regards distribution of pigment on the body.

DUTCH-MARKED SERIES (SERIES D).

The data for inheritance of spots in the Dutch-marked series are contained in tables 5 to 14. The individuals are arranged in a series of groups, each group having a larger average amount of Dutch-marked ancestry than the foregoing one. For convenience in description, we may use the symbol D to designate any Dutch-marked animal, followed by a numeral to indicate the amount in generations of its Dutch-marked ancestry. On this basis an original Dutch-marked animal, one without Dutch-marked ancestors, would be designated D_0; a Dutch-marked animal with one generation of Dutch-marked ancestry, D_1; one with two generations of Dutch-marked ancestry, D_2; and one with three generations of Dutch-marked ancestry, D_3. Similarly a Dutch-marked animal, one parent of which was Dutch-marked, but the other not, might be designated $D_{\frac{1}{2}}$; the Dutch-marked young of $D_{\frac{1}{2}}$ individuals might be called $D_{1\frac{1}{2}}$; etc.

The young of our Dutch-marked animals have been divided into four groups, the *average* amounts of Dutch-marked ancestry in these groups being respectively $\frac{1}{2}$, 1, 2, and 3 generations. Groups of individuals were at first created which came midway between the groups mentioned, but these differed so little from the adjacent groups that we decided finally to abolish them, dividing them equally between the groups to which they were intermediate. Thus a group, $D_{1\frac{1}{2}}$, was divided equally between the groups D_1 and D_2, and a group, $D_{2\frac{1}{2}}$, between the groups D_2 and D_3.

For each individual observed, a record was made of what spots it possessed and what spots its parents possessed. These records are shown in tables 5, 7, 9, and 11 for the four groups of individuals having the mean ancestral values $D_{\frac{1}{2}}$, D_1, D_2, and D_3 respectively. Each table contains separate records for head, shoulder, side, and rump spots. The significance of the various entries will be sufficiently clear after an examination of a single category (head spots) in table 5. There are for the individuals examined three possible conditions as regards head spots. An individual may have two such spots (one on either side of the head); it may have only one head spot (right or left, as the case may be); or it may have no head spots. Records are made for these three conditions separately in the horizontal columns marked 2 spots, 1 spot, and 0 spots respectively. For the possible head-spot conditions of the two parents considered jointly, five categories are necessary, 4, 3, 2, 1, and 0. These are indicated in vertical columns in the tables. An entry in the upper left-hand square of the table signifies an individual having a pair of head spots, born of parents both of which were similarly marked; an entry in the next square to the right means an individual having likewise a pair of head spots, but born of parents only one of which had a pair of head spots, the other having one head spot only. The table, accordingly, it will be observed, represents the summation of a large amount of detailed information concerning the distribution of spots in offspring and parents, respectively.

For convenience in comparing the categories of individuals one with another, the entries in the tables are expressed in per cents, as well as in numbers of cases observed. Thus in table 5, left-hand vertical column, entries are made concerning the head-spot conditions of 49 young born of parents both of which had two head spots. Of these 49 young, 42, or 85.7 per cent, had two head spots, as each of the parents had; 6, or 12.3 per cent, had only one head spot; while 1 individual, or 2 per cent, had no head spots. The other columns of the table are to be interpreted in a similar way.

Glancing over the four divisions of table 5, which includes observations on 117 young, we notice that the offspring of parents both of which possess a particular pair of spots (left-hand vertical column) possess that same pair of spots in from 70 to 86 per cent of all cases. The agreement between parents and offspring is greatest (85.7 per cent) in the case of head spots, and least (70 per cent) in the case of side spots.

From the last vertical column of table 5 we learn that when both parents lack shoulder or side spots, 50 to 58 per cent of their young lack the corresponding pair of spots entirely.

These facts seem to indicate a rather strong degree of inheritance of particular spots upon the body of the guinea-pig; so we at first interpreted them, but more extended observations limit this conclusion considerably.

The four divisions of table 5 are combined in table 6, which accordingly shows, for the body as a whole, to what extent particular spots are inherited in the group of 117 individuals with which tables 5 and 6 deal. Inheritance of paired spots is seen to be, on the average, measured by 78.8 per cent (left-hand upper square, table 6); absence of paired spots is seen to be measured in inheritance by 55 per cent (lower right-hand square, table 6). Further, the fewer spots the parents have, the fewer their young have, as a glance at the horizontal columns of table 6 will show. This is true not only of the body as a whole (table 6), but also of its individual parts (table 5). *Particular amounts* of pigmentation are accordingly inherited quite strongly; distribution of that pigment in certain of the typically pigmented areas, rather than in others, is really not inherited, as we shall see, though such is the impression first given by these tables.

For the second group of individuals of Dutch-marked ancestry (mean ancestral condition, D_1), the observations concerning spot inheritance are given in tables 7 and 8; for the third group (D_2), in tables 9 and 10; and for the fourth group (D_3), in tables 11 and 12. The numbers of individuals in these several groups are 30, 178, and 136, respectively, making in the Dutch-marked series as a whole 461 individuals observed.

To obtain averages for the entire series, with which the values for the several groups might be compared, tables 5, 7, 9, and 11 have been combined to form table 13, which accordingly shows the *average* degree of inheritance, in the entire series, of head spots, shoulder spots, side spots, and rump spots, respectively. Similarly, tables 6, 8, 10, and 12 have here been combined in table 14, which accordingly shows *in general* to what extent paired spots are inherited in the Dutch-marked series.

A comparison of the inheritance coefficients (per cents) in corresponding places of tables 5, 7, 9, and 11 shows remarkably little change from one generation to another. Paired head spots are inherited in the four groups in 85.7, 82.1, 86.5, and 89.3 per cent of the cases, respectively, or in the series as a whole, in 87 per cent of all cases (table 13, upper left-hand square). Similarly, paired rump spots are inherited in the four groups in 71.4, 80, 80.9, and 83.1 per cent of the cases, respectively, or in the series as a whole (table 13) in 80.7 per cent of all cases. In both head and rump spots the inheritance coefficients increase slightly with the progress of selection.

Shoulder and side spots are so frequently lacking in the Dutch-marked series that comparison of the per cents for presence of these pairs would have little value; it is better to examine rather the per cents for *absence* of these spots. Both shoulder spots are absent in the four groups in 58.5, 59, 41, and 42 per cent of the cases, respectively (a *decreasing* series); average for the four groups, 46.3 per cent. Similarly, side spots are entirely wanting in the several groups in 50, 34.7, 47.2, and 44.8 per cent, respectively (again, on the whole, a decrease), the average for the four groups being 46 per cent.

It is scarcely necessary to examine in detail other parts of the several tables. They show conditions for the most part intermediate between those seen in these extreme instances.

Selection for presence of spots (head and rump) and for absence of spots (shoulder) seems, in the same series of observations, to give contrary results. These seemingly inconsistent facts are really not inconsistent. Spots desired present and selected *for* have become slightly oftener present; spots desired *absent* and selected *against* have likewise become slightly oftener *present*. Selection for particular spots has had nothing to do with bringing about the result observed. The pigmentation has become somewhat more extensive in the race, without our having consciously aimed to make it so. Spots occur somewhat more numerously in the later than in the earlier generations of the series (see table 26). The increase, however, takes place no oftener in regions where increase is desired (head and rump) than where it is *not* desired (shoulder), or where its occurrence is a matter of indifference (side).

This result shows the entire inefficiency of selection in guinea-pigs to fix a coat-pattern which is dependent upon the distribution of pigment in particular spots upon the body.

Head spots regularly give higher higher inheritance coefficients than any other
t of spots, a possible reason being that head spots really include two
istinct pairs of pigment areas, namely, eye spots and ear spots. Commonly,
 the Dutch-marked series, both of these are pigmented, in which case it
akes no difference whether the two are dealt with separately or jointly·
ut sometimes an eye area may be pigmented, while the adjacent ear area
unpigmented, or *vice versa;* in such cases separate enumerations of eye
nd of ear spots would indicate a less amount of head pigmentation than
e joint reckoning followed, for purposes of convenience, in this paper.
side from this consideration, however, it is certain that the head region is
tener pigmented than any other part of the body.

HEAD-SPOT SERIES (SERIES H).

The data concerning spot inheritance in the head-spot series are contained
 tables 15 to 24. The entries are made separately for the different regions
 the body in tables 15, 17, 19, and 21, for four different groups of indi-
iduals possessing different average amounts of head-spot ancestry. The
verage amounts of head-spot ancestry in these several groups are a half-
eneration, one generation, two generations, and three generations, respec-
vely, and may be expressed by the symbols $H_{\frac{1}{2}}$, H_1, H_2, and H_3, as explained
 connection with the Dutch-marked series. In the group $H_{\frac{1}{2}}$ (table 15),
e of the parents of each individual recorded bore certain spots other than
ead spots, but in the other three groups of Series H (tables 17, 19, and 21),
one of the parents bore spots other than head spots. Tables 15, 17, 19,
nd 21 are combined in table 23, which shows the average degree of in-
eritance of particular spots in the series as a whole.

As in the case of Series D, the entries for the different regions of the body
ave, in the case of each group of individuals, been combined to get an expres-
on for spot inheritance *in general*, in the particular group of individuals
nder consideration. These combination entries are found in tables 16, 18,
o, and 22, for the four groups considered separately, and combined into
ne table for the whole head-spot series, in table 24.

In this series, rigid selection has been exercised *for* head spots, and *against*
ll other spots. We will now consider what the effects of this two-fold
election are. Paired head spots are present in the following per cents of
he young of parents possessing such spots, in the several groups: 79.8,
5.4, 77.1, and 82.1 per cent; average for the series (table 23), 81.1 per cent.*
o change is observed uniformly in the direction either of increase or of

---

*In obtaining the average, each table is weighted in proportion to the number of individ-
als recorded in it. These are, for the successive groups, 207, 193, 129, and 58; total, 587.

decrease of head spots, but they are on the whole less common in this series than in Series D (81.1 and 87 per cent, respectively).

However, shoulder, side, and rump spots become steadily less common in the series, as a glance at table 25 will show. It is doubtful, however, whether this is to be regarded as an effect of selection for elimination of specific areas It indicates, rather, a decrease in the average *amount* of pigmentation possessed by animals of this series, unwittingly brought about by the attempt to eliminate most of the spots typically pigmented (compare table 27). It is significant that head spots, though desired present and chosen equally in Series D and in Series H, occur more commonly in the former than in the latter. Animals of Series D are *more extensively* pigmented than those of Series H (compare tables 26 and 27); the additional pigmentation may occur in any of the typically pigmented areas, regardless of the areas which happen to have been pigmented in the selected ancestors.

Nevertheless, we find a more rapid falling off in the pigmentation of some areas than in that of others. In Series H, shoulder spots fall off 16.9 per cent; side spots, 7.3 per cent; and rump spots, 6.4 per cent; while head spots, as already noted, show little change. Now, selection, it will be remembered, was exercised for *elimination* of shoulder, side, and rump spots alike, but for the *retention* of head spots. Accordingly we do not get changes uniformly consistent with selection. Regardless of selection for specific spots, we get, when the total amount of the pigmentation is decreasing, a decrease most rapidly in shoulder spots, less rapidly in side spots, less rapidly still in rump spots, while the head spots are least affected. Conversely, in a series in which the amount of pigmentation is increasing, the increase is less marked in head than in rump spots, and less in side than in shoulder spots. Thus, in Series D, head spots increase 3.6 per cent, rump spots, 11.7 per cent; the change in side spots (entire absence) is 6.2 per cent, of shoulder spots, 16.5 per cent.

In further support of this same idea—that with increasing or decreasing pigmentation changes occur more rapidly in some areas than in others, irrespective of the *particular* spots borne by the ancestors—we may compare the averages for Series D and H (tables 13 and 23). The differences between the two series are, in frequencies of head spots, 5.9 per cent; of side spots (absent), 20 per cent; of shoulder spots (absent), 25.4 per cent. As regards rump spots, no comparable entries occur in either the first or the last columns of tables 13 and 23, but an examination of the middle column (2) in both tables indicates that the difference in rump spots in the two series is about as great as in head spots.

These facts are in harmony with an observation made in advance of the statistical investigation, that the most persistent of all the pigmented areas are the head spots, and next in order of persistency are the rump spots.

Our general conclusion from a study of these two series of guinea-pigs, including 1,048 individuals, is that one can by selection either increase or decrease the extent of the pigmented areas, but it is impossible by selection to fix this pigmentation in a particular pattern, retaining pigment areas on certain parts of the body and eliminating them from others. As the pigmentation changes in extent, under the influence of selection, the various areas typically pigmented are affected in the following order: Shoulder, side, rump, and head, the change being greatest in the first-named and least in the last-named area, irrespective of what particular spots were present in the selected ancestors.

STATISTICAL ANALYSIS OF THE DATA FOR SERIES D AND H.

The foregoing conclusions may, we think, fairly be drawn from an examination of the tables as they stand. For those, however, who place confidence in the more precise methods of statistical analysis devised by Pearson and others, it may be more satisfactory to treat the tables, which have been constructed for the various groups of individuals, as *correlation tables*, and derive from them the constants which measure the variability of parents and children respectively in the several groups, and the degree of correlation between the two. Such constants are given, in tables 26 and 27, for the several groups of Series D and H, respectively.

From the left-hand columns of table 26 we learn that, in Series D, the number of spots borne by an individual increases from group to group, in the case of both parents and offspring; but the number of spots is, in every case, 20 to 26 per cent greater in the offspring than in their parents. This indicates a tendency for the offspring to become pigmented in regions which lacked pigment in their selected parents.

Table 27 shows the existence of a similar, but still stronger, tendency in Series H, the offspring bearing 50 to 105 per cent more spots than their parents. This tendency may be considered regression, a tendency to return to a condition of more *widely distributed* pigmentation, and to acquire spots where the selected ancestors lacked them. It does not imply that the young bear *more* pigment than their parents. The regression is stronger the more reduced the pigmentation of the parents, as we might expect. This is seen to be true both within Series H and in that series as a whole compared with Series D.

The standard deviation from the average number of spots is for the offspring a very constant quantity, being close to 3.6 in both Series D and Series H. This indicates no change in the variability in number of spots as a result of selection, or as a consequence of change in the number of pigmented areas. The parents show throughout Series H and in Group $D_{\frac{1}{2}}$ of Series D a less standard deviation than their offspring. This fact, however, has for our purposes little significance, since the parents are *selected individuals*, while their offspring are not. The same may be said of the coefficients of variability for the parents. For the offspring, the coefficient of variability (ratio of standard deviation to mean, times 100) decreases slightly in Series D, and increases in Series H. This must not be interpreted as signifying a change in variability of an opposite nature in the two series. It is due entirely to the changes in the mean (average number of spots), these changes being opposite in nature in the two series. If the *mean* remained constant throughout each series, the amount of variability indicated by the coefficient of variability would likewise remain constant; for we have seen that the standard deviation is constant, irrespective of changes in the mean and irrespective of the number of generations during which selection has been in progress. Standard deviation is, therefore, a better measure of variability than the coefficient of variability in these series.

The coefficient of correlation (r) has little real significance in tables 26 and 27. This constant is relatively small in the Dutch-marked series and grows smaller as selection progresses, whereas in the head-spot series it is relatively large and grows larger with the progress of selection. Selection does not have an opposite effect in the two series upon the inheritance of a coat-pattern. The whole effect is due to change in the *amount* of the pigmentation, not to its distribution. When the amount of pigmented surface is small, its distribution in certain spots (head and rump) is more certain and the *correlation rises* (Series H). When, on the other hand, the pigmented surface is large (Series D, table 26), its distribution is less certain and the correlation is low; it becomes lower as the pigmentation increases, in spite of progressive selection.

The statistical analysis confirms the conclusions previously drawn from our observations. Selection is powerless to fix a particular coat-pattern not dependent upon *amount* of pigmentation. It is as powerless to decrease the variability in number of spots as to fix any pattern formed by them.

NOSE-SPOT SERIES (SERIES N).

It is only in the case of the nose-spot series (Series N) that we are able to detect influence of selection in fixing a pigment-pattern among guinea-pigs. Two individuals showing this rather striking variation in pattern were figured by Castle (:05, pl. 6); another is shown in pl. 2, fig. 1, of this paper. The starting-point of the series of selections for the fixation of this pattern was the male 1989^a shown in fig. 12, pl. 6 (Castle, :05). Besides the nose spot shown in the figure, this animal bore a distinct right shoulder spot, not visible in the figure. This animal produced a considerable number of young with nose-spot markings similar to his own, as will be seen from an examination of table 28. As a result of matings with animals which did not bear rose spots, he produced four individuals with nose spots and fourteen without nose spots, or 22.2 per cent of nose-spot (NS) young. When he was mated with these nose-spot young ($NS_{\frac{1}{2}}$) or with nose-spot individuals produced in other experiments but not of nose-spot ancestry, and so designated NS_0, he produced a much higher percentage of nose-spot young, viz, 53.6 per cent by NS_0 mothers, and 61.5 per cent by $NS_{\frac{1}{2}}$ mothers. In one case he was mated with his NS_1 daughter by an NS_0 mother, from which mating there resulted three nose-spot young and two otherwise marked, or 60 per cent NS. The number of young in this last experiment is too small to be very significant quantitatively, but those in the other matings are large enough. They show an increase in the per cent of NS young with an increase in the amount of NS ancestry, and strongly suggested the possibility of a still further increase by continued selection. Such increase, however, has not up to this time been realized. Five sons or grandsons of ♂1989[a] have been quite extensively tested and the same is true of one individual (♂5595, NS_1, table 29) descended from a brother of 1989[a]. Of these six males, two have records about equaling in production of nose-spot young the record of ♂1989[a], but the records of the remaining four are inferior to his. The two males with the best records are ♂5652 (table 29), a son of ♂1989[a], and ♂5669 (table 29), a grandson of ♂1989[a] and son of ♂5151 (table 29). Neither of these males, however, has produced a large number of young (the former 16, the latter 10), and so too much importance must not be attached to the per cent results. Of all the six males, 5151 alone has produced more than fifty young. By NS_0 and NS_1 mothers alike, about 40 per cent of his young bear nose spots. His father's record by the same group of mothers was about 15 per cent better (see table 28).

The collective results given by the six males enumerated in table 29 are shown in table 30. They indicate that in the long run a higher percentage of NS young is produced by NS_1 than by NS_0 sires, but neither group gives as high a percentage as the original nose-spot male, 1989[a] (table 28).

The nose-spot mothers employed in these experiments are much more numerous than the males, but because the number of young produced by any one of them is in no case greater than fifteen, the results are not given for the mothers individually, but only in collective form (table 30, last part).

The collective results for the females, like those for the males, show an increase in the percentage of nose-spot young with increase in the amount of nose-spot ancestry. In other words, the character is apparently inherited feebly through both the male and the female lines, and is being gradually fixed by selection. The process, however, is a slow one. After two generations of selection, the inheritance coefficient is no higher than in the original male, 1989[a]. Whether it can ultimately be made higher remains an open question.

The objection may be made that a nose spot as such is not inherited, any more than an eye spot or a rump spot (see page 25); that in a spotted race a certain average amount of pigmentation is an inherited condition, but the distribution of this pigment is wholly a matter of chance. To test this point, comparison has been made with the young of certain sires in Series *H* (table 31). In the case of the litters recorded in this table, neither parent bore a nose spot, nor came from a nose-spot family. The extent of the pigmentation of the parents is measured roughly by the *number* of the typical spot-areas which were pigmented (see columns 2, 3, 4, 7, and 10, table 31). For comparison with the nose-spot series, see columns 3 and 6 of tables 28, 29, and 30. It will be observed that the *extent* of the pigmentation is similar in the nose-spot and head-spot series, being in both cases close to an average of three spots to an individual. The percentage of nose-spot young, however, is much lower in the eye-spot than in the nose-spot series, being 16.3 per cent in the former as compared with 40 to 50 per cent in the latter. That the nose-spot marking does occur quite frequently among spotted guinea-pigs not selected for that character is shown clearly by table 31. That its occurrence, however, is *more* frequent when selection *is* made for the character is evident from tables 28 to 30.

In table 31, as in tables 28 to 30, it will be observed that nose-spot young have fewer *other* spots than do their brothers and sisters that lack nose spots. The average difference is from half to two-thirds of a spot. This means that a nose spot takes the place, to some extent, of pigmentation elsewhere, and it is in part a matter of chance whether the pigmentation is located on the nose or elsewhere. But chance is not the *only* element (if we may so speak of chance) entering into the matter, for nose-spot parents clearly produce *more* nose-spot young than do other parents transmitting a like amount of pigmentation. How, then, are we to account for the fact indicated by our observations that, while other patterns are unfixable, the nose-spot pattern is in part at least fixable? At present we can not account for it, but a consideration of familiar facts concerning mammalian development may help us in shaping a hypothesis.

The production of hair and skin pigments in guinea-pigs is the exclusive function of the ectoderm, as shown by Leo Loeb ('97) and confirmed by Castle (:05). In spotted guinea-pigs the limits of the pigment spots are very precisely defined at birth and these limits, so far as we have been able to observe, are never subsequently transgressed in the slightest degree. Areas which are white at birth remain white ever afterward;* areas which

---

*We leave out of consideration for the present the "peripheral" pigmentation which albinos as well as spotted animals may possess (see Castle, :05). This is *not* fully developed at birth.

are black at birth remain black; those which are yellow remain yellow; and those which contain at birth black and yellow hairs interspersed remain in that condition ever afterward. This indicates that every portion of the epidermis has its pigment-forming capacity early and finally differentiated. In spotted animals the capacity to form hair and skin pigments is transmitted only to certain portions of the epidermis. Our statistical studies make it clear that guinea-pigs and rats (and probably other spotted mammals also) transmit with a good deal of constancy definite amounts of pigmented surface, but that, in guinea-pigs at least, the *distribution* of this pigmentation over the body is not strictly localized in the germ. A certain amount of pigment, apparently, is handed over to the epidermis, but it seems to be to some extent a matter of chance upon what part of the body this pigmented epidermis finally comes to lie. It is not, however, *entirely* a matter of chance. It is almost certain that the pigment will lie chiefly on the dorsal surface, and if the pigmentation is not extensive, it will be restricted to one or more of the regions which we have designated *eye, ear, shoulder, side,* and *rump* areas, all of which are paired and frequently separated one from another by intermediate unpigmented areas. In other cases, even when adjacent areas are confluent, they show their essential distinctness by sharp differences in color. Besides the paired areas mentioned, there is an unpaired area at either end of the body. The anterior one we have designated nose spot, the posterior one might be called a tail spot, though in the guinea-pig it is scarcely distinguishable from the rump spots, because there is no external tail. The distinct pigment spots are derivatives, doubtless, of individual blastomeres set apart early in development for the production of the epidermis. We know that in birds and mammals the epidermis is first differentiated along either side of the primitive streak. The ectoderm along the middle of the primitive streak sinks down to form the neural canal, then the divided right and left halves of the epidermis come together above it, while anterior and posterior to the neural invagination the right and left halves of the epidermis have been from the beginning continuous. Very likely the nose- and tail-spot regions correspond with these regions of original continuity of the right and left halves of the epidermis, while the paired areas are formed out of the epidermis lateral to the neural invagination. The epidermis of the ventral side of a bird or mammal is developed, we know, later than the dorsal portions, and in spotted individuals, apparently, the amount of pigment inherited is insufficient ordinarily to extend out over the blastoderm into this region. Accordingly the pigment is restricted to the portions of the epidermis first differentiated, *i. e.*, adjacent to the primitive streak. In most cases, it is insufficient to cover more than portions even of that.

On this interpretation, the difference between a black-eyed white and a pink-eyed white (or albino) guinea-pig is this: The neural plate of the former receives a pigment potentiality from the ectoderm which is invaginated to form the neural canal (out of which the retina of the eye is later formed); the albino does not receive the pigment potentiality into the neural plate, simply because that potentiality is absent from the entire ectoderm. In the black-eyed white guinea-pig no part of the epidermis is pigmented, because all the color-bearing ectoderm has gone into the neural invagination and none is left outside for the epidermis. When a very little is left outside, it usually is found either upon an ear, upon one side of the forehead, or about the eye (see Castle : 05), facts which indicate with considerable clearness what relation the epidermis of the head region bore to the anterior part of the neural invagination, and support the view that the nose spot arises out of epidermis originally *anterior* to the neural plate. Black-eyed white guinea-pigs *usually* produce offspring with one or more spots on the body, *i. e.*, the amount of pigment which they transmit is commonly large enough to extend lateral or anterior to the neural plate and cause epidermal spots as well as pigmented eyes. Thus a pair of black-eyed white guinea-pigs has recently (Nov., 1906) produced one black-eyed white young and two with nose spots and an ear spot each. In spotted guinea-pigs the pigment is apparently always *centralized* about the primitive streak. It invariably passes into the neural canal (if any pigment* is inherited at all), so that the eyes are pigmented; usually it extends out also into the epidermis bordering on the neural plate, but its limits are variable. When the amount of epidermal pigment is small, it is found most often upon the head, adjacent to the portion of the neural plate out of which the eyes were formed.

If we could follow through the period of cleavage the history of the pigment-forming substance of the fertilized egg, we should probably find that in spotted guinea-pigs this is distributed to certain blastomeres in the animal hemisphere of the egg. Apparently it is to some extent a matter of chance *what* blastomeres receive the pigment, but the anterior part of the neural invagination is *sure* to receive pigment, and the adjacent portions of the skin are *more* likely than any other regions to do so. In nose-spot animals, pigment is evidently distributed to the most anterior epidermal cells which take part in forming the dorsal surface of the animal. If, as our observations indicate, the condition is to some extent hereditary, the pigment potentiality must be localized in both egg and sperm in that

---

*See footnote, page 28.

part of the germ-cell which gives rise to the anterior part of the primitive streak. In selecting for the nose-spot character, accordingly, we have simply selected for *extreme anterior localization* of a reduced amount of pigmentation. Selection for localization less far forward is apparently less effectual, probably because, in that part of the germ, a cell-division, which would involve unequal distribution of the pigment potentiality to the daughter cells, might carry the potentiality *either* forward or backward in the embryo, whereas original extreme anterior localization could result in its transportation only forward to the most anterior part of the embryo. This, while a purely hypothetical explanation, is offered as a suggestion to the embryologist, who some day, perhaps, will be able to identify in the germ and trace through its various stages of development the substance or substances on the presence of which pigment formation depends.

If *anterior* localization of pigment is possible through selection, it would seem that *posterior* localization should likewise be possible. We began at one time a series of experiments to test this point, but the results given by the first selected generation were so unpromising that the experiment was abandoned. It may be pointed out, however, that one might expect fixation of such a pattern to be more difficult, because, like the Dutch-marked pattern, it involves a double selection, viz, selection for the posterior localization of pigment sought (tail spot) and simultaneously and unavoidably selection for the anterior localization of pigment represented in the black eyes of all spotted individuals.

The explanation which has been offered for the distribution of pigment in partial-albino guinea-pigs will apply equally well to the case of rats, with this difference: In rats the body is more elongated than in the guinea-pig, and pigment-reduction affects at first rather the lateral than the longitudinal distribution of pigment. The back-stripe first becomes narrower, then becomes interrupted, and finally drops out altogether, leaving pigment only on the head and usually also sparingly on the tail (see pl. 1). This condition corresponds with that condition of the guinea-pig in which head and tail (or rump) spots only are present, a condition very common in Series *H*, as we have seen. As the hood becomes reduced in extent, the white areas first extend forward ventrally to the mouth, then in the median dorsal line between the shoulders, and simultaneously a white spot appears in the forehead (see pl. 1, fig. 3). Further reduction than this has not yet been obtained, but it is evident that the pigmented areas are becoming restricted toward the sides of the head adjacent to the eyes, precisely as in guinea-pigs.

CONCLUSIONS.

The results of selection brought to bear upon the coat-pattern are seemingly very different in rats and in guinea-pigs, yet a careful analysis of the facts shows the results to be not so dissimilar in the two cases as they at first thought appear.

In both rats and guinea-pigs we can by selection increase or decrease at will the average extent of the pigmented areas. In both rats and guinea-pigs the extent of the pigmented areas varies continuously, and out of these continuous variations permanent modification of the pigmentation can be secured.

Reduction in the total amount of the pigmentation is attended in rats by restriction of the pigment to very definite areas, whereas in guinea-pigs it may be distributed in any or in all of a series of spots. Herein lies the whole difference between the two cases. When in rats we select for reduced pigmentation, we get animals with a narrow or interrupted back-stripe and with a less extensive hood; when in guinea-pigs we make a similar selection, we get animals with fewer or less extensive spots. We can not in guinea-pigs decide arbitrarily *which* areas shall be pigmented (except, possibly, in the case of nose spots), any more than in rats we can at the same time increase the extent of the hood and decrease that of the back-stripe.

In rats, we have as a result of pigment reduction a series of coat-patterns, each breeding true within certain limits; in guinea-pigs, the fluctuation in the extent of the pigmented areas is probably no greater than in rats, but because the pigmented areas do not disappear in as definite an order during pigment reduction, we have no constant coat-patterns. Nevertheless there is every reason to suppose that different degrees of pigmentation are inherited in Mendelian fashion in guinea-pigs, precisely as they are in rats. If the pigment reduction followed a definite course in guinea-pigs, as it does in rats, this would be easily recognizable in the coat-pattern. As it is, measurement of the extent of the pigmented areas would be necessary to make it apparent. This we have not undertaken to do in the case of guinea-pigs; we have merely taken account of the regions pigmented, not of their extent. This probably explains in part why regression is observed in the selection experiments with guinea-pigs, but not with uniformity in those with rats. In guinea-pigs we attempted by selection to restrict the *number* of the pigmented areas; this was found to be impossible except as it occurred incidentally to reduction in the total *amount* of pigmentation. The regression occurred in *number* of pigmented areas, not, so far as we know, in the total amount of the pigmentation. We have no doubt, however, that such regression would be found to occur in cases in which extreme variates were selected. We have found it so in selecting black-eyed white guinea-pigs, those with no pigment except in the eye. Almost invariably the young of such animals have borne more pigment than did their parents. A similar result would doubtless follow selection of self-pigmented rats obtained from Irish parents. No doubt many of the young would bear some white fur. With selection of less extreme variates, regression less extreme may possibly occur, though our statistical observations do not show any regression in the case of rats.*

If regression does occur, can we with propriety consider the effects of selection permanent? De Vries has answered this question in the negative on the basis of his selection experiments with maize, striped flowers, double buttercups, and other similar material. It seems to us, however, that the answer should be qualified. The final result will depend upon the amount and the persistency of the regression. In De Vries's experiments with maize, as in those of Fritz Müller ('86), the regression grows less with each selection. If this continued, the regression should ultimately become a negligible quantity. After repeated selection for a desired extreme condition, the race should become stable at a condition only a little less extreme than that selected.

De Vries's fine series of selection experiments with the buttercup (*Ranunculus bulbosus*) seems to the writers scarcely to justify the conclusion that selection has no permanent effects. Starting with a one-sided or "half-Galton" variation curve, with a range from the modal number, 5, upward to 13, De Vries was able by selection for an increased number of petals to raise the mode to 11, the average to 8.6, and the upper limit of variation to 31, and to obtain a two-sided, or Galtonian, variation curve with only a moderate amount of skewness, and with greatly diminished regression. All this was accomplished within five generations.

We consider the selection question still an open one. Further experiments and longer continued ones are needed. Our own observations, so far as they go, and those of Fritz Müller and De Vries, lead us to think that selection is a most important factor, not only in the isolation of discontinuous variations, but also in their production.

Further, we are far from convinced that all evolutionary progress is to be attributed to discontinuous variations, any more than to Mendelian inheritance. The distinction between continuous and discontinuous variations is a useful one, just as that between alternative and blending inheritance, but a sharp line of division can be drawn in neither case. The hooded and Irish coat-patterns of rats are recognized discontinuous variations, alternative in inheritance, yet our lot M of hooded rats is as nearly intermediate between typical hooded and typical Irish rats as anything that can well be imagined. The coat-patterns of fancy rats, though discontinuous as they ordinarily occur, can be transformed into continuous variations. Concerning

---

*April, 1907. In this year's experiments we are getting some evidence of the expected regression.

the hooded and Irish patterns, Doncaster (:06), after an extended experience, says (p. 216): "Only once have I had any hesitation in classing a rat as belonging to one or the other." Yet we have seen that by cross-breeding and selection these same discontinuous groups can be made continuous by the production of any desired number of intermediate groups, each varying continuously about a different mode.

Again, though the inheritance is clearly Mendelian, when hooded and Irish rats are crossed, the gametes formed by cross-breds are not pure, but modified, each extracted pattern being changed somewhat in the direction of that pattern with which it was associated in the cross-bred parent. This means simply that the inheritance, though in the main alternative, is to some extent blending.

Since it is impossible to make a sharp distinction between continuous and discontinuous variations, as well as between blending and alternative inheritance, it is fallacious to assign all evolutionary progress to one sort of variation or to one sort of inheritance.

TABLES.

TABLE 1.—*Actually different classes into which we may expect the five visibly different classes of albino and partial-albino rats to fall.*

[Twenty-two of the twenty-seven classes enumerated have been proved to exist.]

	Classes visibly different.				
	Gray Irish.	Gray hooded.	Black Irish.	Black hooded.	Albino.
Classes expected	1 GI 2 GI(W) 2 GI(B) 2 GI(H) 4 GI(W.B) 4 GI(W.H) 4 GI(BH) 8 GI(W.BH) ...	1 GH 2 GH(W) 2 GH(B) 4 GH(W.B)	1 BI 2 BI(W) 2 BI(H) 4 BI(W.H)	1 BH 2 BH(W)	1 W[BH] 1 W[BI] 1 W[GH] 1 W[GI] 2 W[B-IH] 2 W[G-IH] 2 W[GB-H] 2 W[GB-I] 4 W[GB-IH]
Total ...	27	9	9	3	16

B, black; G, gray; H, hooded; I, Irish; W, total albinism. Symbols indicating unseen recessive characters are placed within parentheses ( ), those indicating characters latent in albinos are placed within brackets [ ].

When two symbols only are used the first refers to color, the second to coat-pattern.

When more than two symbols stand together within brackets, those which refer to color are placed at the left of a hyphen, those referring to pattern at its right.

Total albinism is indicated by W. When albinism is recessive with other characters in pigmented individuals, the W will be separated by a period from the symbols designating the other characters.

The numerals prefixed to the several class-designations indicate the expected frequencies of the classes when individuals are mated *inter se* which have the characters indicated by the designation GI (W.BH).

TABLE 2.—*Expected and observed distribution of young produced by partial-albino rats mated inter se.*
[Abbreviations as in table 1.]

Mating No.	Nature of mating.	Proportions of offspring expected on Mendelian hypothesis.					Observed result (expected result, italic).				
		GI	GH	BI.	BH.	W.	GI.	GH.	BI.	BH.	W.
1	GI(W.BH) bred inter se..	27	9	9	3	16	11 / *7.2*	2 / *2.4*	1 / *2.4*	1 / *0.8*	2 / *4.2*
2	GI(W.BH)×GH(W.BH)	9	9	3	3	8	8 / *9*	10 / *9*	5 / *3*	2 / *3*	7 / *8*
3	GI(W.BH)×BI(W.BH)..	9	3	9	3	8	8 / *6.2*	1 / *2.1*	4 / *6.2*	4 / *2.1*	5 / *5.5*
4	GI(W.BH)× BH(W)......	3	3	3	3	4	13 / *15*	16 / *15*	13 / *15*	24 / *15*	14 / *20*
5	GI(W.BH)× BH............	1	1	1	1	0	6 / *3.2*	3 / *3.2*	0 / *3.2*	4 / *3.2*	0 / *0*
6	GI(W.B)× BI(H)..........	1	0	1	0	0	3 / *4*	0 / *0*	5 / *4*	0 / *0*	0 / *0*
7	GI(H)×BH(W)............	1	1	0	0	0	8 / *9*	10 / *9*	0 / *0*	0 / *0*	0 / *0*
8	GH(W.B)×GH(B)........	0	3	0	1	0	0 / *0*	18 / *21*	0 / *0*	10 / *7*	0 / *0*
9	GH(W.B) bred inter se...	0	9	0	3	4	0 / *0*	14 / *12.4*	0 / *0*	2 / *4.1*	6 / *5.5*
10	GH(W.B)×BH(W)........	0	3	0	3	2	0 / *0*	14 / *13.9*	0 / *0*	12 / *13.9*	11 / *9.2*
11	GH(W.B)× BH............	0	1	0	1	0	0 / *0*	4 / *3.5*	0 / *0*	3 / *3.5*	0 / *0*
12	BI(W.H) bred inter se....	0	0	9	3	4	0 / *0*	0 / *0*	26 / *27.6*	11 / *9.2*	12 / *12.2*
13	BI(W.H)× BH(W).........	0	0	3	3	2	0 / *0*	0 / *0*	15 / *18*	20 / *18*	13 / *12*
14	BI(H)× BH.................	0	0	1	1	0	0 / *0*	0 / *0*	5 / *7*	9 / *7*	0 / *0*
	Totals, observed............... Totals, expected..						57 / *53.6*	92 / *91.5*	74 / *86.4*	102 / *86.8*	70 / *76.7*

36　INHERITANCE OF COAT-PIGMENTS AND COAT-PATTERNS

TABLE 3.—*Expected and observed distribution of young produced by partial-albino rats mated with albinos.*
[Abbreviations as in table 1.]

Mating No.	Nature of mating.	Proportions of offspring expected on Mendelian hypothesis.					Observed result (expected result, italic).				
		GI.	GH.	BI.	BH.	W.	GI.	GH.	BI.	BH.	W.
15	GI(W.BH)×W[BH]	1	1	1	1	4	7 / *8.1*	7 / *8.1*	6 / *8.1*	11 / *8.1*	34 / *32.5*
16	GH(W.B)×W[BH]	0	1	0	1	2	0 / *0*	4 / *6*	0 / *0*	6 / *6*	14 / *12*
17	GH(B)×W[BH]	0	1	0	1	0	0 / *0*	11 / *12.5*	0 / *0*	14 / *12.5*	0 / *0*
18	BH×W[GB-H]	0	1	0	1	0	0 / *0*	8 / *11.5*	0 / *0*	15 / *11.5*	0 / *0*
19	BH×W[B-IH]	0	0	1	1	0	0 / *0*	0 / *0*	9 / *11*	13 / *11*	0 / *0*
20	BI(W.H)×W[GB-IH]	3	3	1	1	8	5 / *4.1*	2 / *1.4*	5 / *4.1*	2 / *1.4*	8 / *11*
21	BI(W.H)×W[B-III]	0	0	3	1	4	0 / *0*	0 / *0*	12 / *10.5*	6 / *3.5*	10 / *14*
22	BH(W)×W[GH]	0	1	0	0	1	0 / *0*	8 / *6.5*	0 / *0*	0 / *0*	5 / *6.5*
23	BH(W)×W[B-IH]	0	0	1	1	2	0 / *0*	0 / *0*	24 / *17*	17 / *17*	27 / *34*
	Totals, observed						12	40	56	84	98
	Totals, expected						*12.2*	*46*	*50.7*	*71*	*110*

TABLE 4.—*Average grade (size of stripe) in the various lots of hooded rats of selected parentage, as compared with the average grade of their parents.*

Lot.	No. Individuals.	Average grade.	Average grade of parents.	Deviation of average filial from average parental grade.
E	132	21.1	21.8	−0.7
F	89	12.9	17.8	−4.9
G	83	9.6	11.0	−1.4
H	61	5.6	7.8	−2.2
I	34	4.0	1.7	+2.3
J	111	15.8	12.0	+3.8
K	70	22.2	23.0	−0.8
L	64	36.2	40.7	−4.5
M	19	52.4	50.0	+2.4

IN RATS AND GUINEA-PIGS. 37

TABLE 5.—*Relation between distribution of pigment spots in 117 guinea-pigs of group $D_{\frac{1}{4}}$, and in their parents.*

Offspring.	Parents.										Location of spots and total number of cases.
	4 spots.		3 spots.		2 spots.		1 spot.		0 spots.		
	Cases.	P. ct.	Cases.	P. ct.	Cases.	P. ct.	Cases.	P. ct.	Cases.	P. ct.	
2 spots.	42	85.7	10	55.6	35	70	...	...	...	...	Head; total cases, 117.
1 spot..	6	12.3	8	44.4	15	30	...	...	...	...	
0 spots.	1	2	...	...	...	...	...	...	...	...	
Total..	49	...	18	...	50	...	...	...	...	...	
2 spots.	4	80	3	37.5	8	36.4	...	...	13	15.9	Shoulder; total cases, 117.
1 spot..	1	20	3	37.5	4	18.2	...	...	21	25.6	
0 spots.	...	...	2	25	10	45.4	...	...	48	58.5	
Total..	5	...	8	...	22	...	...	...	82	...	
2 spots.	7	70	3	100	16	36.4	3	75	14	25	Side; total cases, 117.
1 spot..	...	...	...	...	10	22.7	1	25	14	25	
0 spots.	3	30	...	...	18	40.9	..	..	28	50	
Total..	10	...	3	...	44	...	4	...	56	...	
2 spots.	25	71.4	...	...	41	50	...	...	...	...	Rump; total cases, 117.
1 spot..	6	17.2	...	...	20	24.4	...	...	...	..	
0 spots.	4	11.4	...	...	21	25.6	...	...	...	...	
Total..	35	...	...	...	82	..	...	...	...	...	

TABLE 6.—*Combination of the four parts of table 5, showing in general the extent to which pigment spots are inherited in group $D_{\frac{1}{4}}$, 117 individuals (468 observations).*

Offspring.	Parents.										Total number of cases.
	4 spots.		3 spots.		2 spots.		1 spot.		0 spots.		
	Cases.	P. ct.	Cases.	P. ct.	Cases.	P. ct.	Cases.	P. ct.	Cases.	P. ct.	
2 spots...	78	78.8	16	55.2	100	50.5	3	...	27	19.6	468
1 spot ...	13	13.2	11	37.9	49	24.7	1	...	35	25.4	
0 spots...	8	8.0	2	6.9	49	24.7	...	...	76	55.0	
Total...	99	...	29	...	198	...	4	...	138	...	

INHERITANCE OF COAT-PIGMENTS AND COAT-PATTERNS

TABLE 7.—*Relation between distribution of pigment spots in 30 guinea-pigs of group D_1, and in their parents.*

Offspring.	Parents.									Location of spots and total number of cases.
	4 spots.		3 spots.		2 spots.		1 spot.		0 spots.	
	Cases.	P. ct.	Cases.	P. ct.	Cases.	P. ct.	Cases.	P. ct.	Cases.	P. ct.	
2 spots.	23	82.1	1	...	...	...	...	...	...	...	Head; total cases, 30.
1 spot..	5	17.9	1	...	...	...	...	...	...	...	
0 spots.	...	...	...	...	...	...	...	...	...	...	
Total..	28	...	2	...	...	...	...	...	...	...	
2 spots.	...	...	2	...	...	...	1	...	4	18.3	Shoulder; total cases, 30.
1 spot..	...	...	2	...	...	...	1	...	5	22.7	
0 spots.	...	...	1	...	...	...	1	...	13	59.0	
Total..	...	...	5	...	...	...	3	...	22	...	
2 spots.	...	...	...	...	2	...	...	...	13	50.0	Side; total cases, 30.
1 spot..	...	...	...	...	2	...	...	...	4	15.3	
0 spots.	...	...	...	...	...	...	...	...	9	34.7	
Total..	...	...	...	...	4	...	...	...	26	...	
2 spots.	24	80	...	...	...	...	...	...	...	...	Rump; total cases, 30.
1 spot..	3	10	...	...	...	...	...	...	...	...	
0 spots.	3	10	...	...	...	...	...	...	...	...	
Total..	30	...	...	...	...	...	...	...	...	...	

TABLE 8.—*Combination of the four parts of table 7, showing spot inheritance in group D_1, 30 individuals (120 observations).*

Offspring.	Parents.									Total number of cases.
	4 spots.		3 spots.		2 spots.		1 spot.		0 spots.	
	Cases.	P. ct.	Cases.	P. ct.	Cases.	P. ct.	Cases.	P. ct.	Cases.	P. ct.	
2 spots...	47	81.0	3	...	2	...	1	...	17	35.4	120
1 spot....	8	13.8	3	...	2	...	1	...	9	18.8	
0 spots...	3	5.2	1	...	...	...	1	...	22	45.8	
Total....	58	...	7	...	4	...	3	...	48	...	

IN RATS AND GUINEA-PIGS. 39

TABLE 9.—*Relation between distribution of pigment spots in 178 guinea-pigs of group D2, and in their parents.*

Offspring.	Parents.									Location of spots and total number of cases.
	4 spots.		3 spots.		2 spots.		1 spot.		0 spots.	
	Cases.	P. ct.	Cases.	P. ct.	Cases.	P. ct.	Cases.	P. ct.	Cases.	P. ct.	
2 spots.	154	86.5	...	...	...	...	...	...	...	...	Head; total cases, 178.
1 spot..	24	13.5	...	...	...	...	...	...	...	...	
0 spots.	...	...	...	...	...	...	...	...	...	...	
Total.	178	...	...	...	...	...	...	...	...	...	
2 spots.	...	...	3	...	5	26.3	7	46.7	49	35.3	Shoulder; total cases, 178.
1 spot..	...	...	1	..	7	36.8	4	26.7	33	23.7	
0 spots.	...	...	1	...	7	36.8	4	26.7	57	41.	
Total.	...	...	5	...	19	...	15	...	139	...	
2 spots.	6	54.5	4	...	22	53.6	5	41.7	39	36.1	Side; total cases, 178.
1 spot..	2	18.2	2	...	7	17.1	1	8.3	18	16.7	
0 spots.	3	27.3	...	...	12	29.3	6	50.	51	47.2	
Total.	11	...	6	...	41	...	12	...	108	...	
2 spots.	144	80.9	...	...	...	...	...	...	...	...	Rump; total cases, 178.
1 spot..	16	9.0	...	...	...	...	...	...	...	...	
0 spots.	18	10.1	...	...	...	...	...	...	...	...	
Total.	178	...	...	...	...	...	...	...	...	...	

TABLE 10.—*Combination of the four parts of table 9, showing spot inheritance in group D2, 178 individuals (712 observations).*

Offspring.	Parents.									Total number of cases.
	4 spots.		3 spots.		2 spots.		1 spot.		0 spots.	
	Cases.	P. ct.	Cases.	P. ct.	Cases.	P. ct.	Cases.	P. ct.	Cases.	P. ct.	
2 spots...	304	82.8	7	63.6	27	45.0	12	44.4	88	35.6	712
1 spot. ..	42	11.4	3	27.3	14	23.3	5	18.5	51	20.7	
0 spots...	21	5.7	1	9.1	19	31.7	10	37.0	108	43.7	
Total...	367	...	11	...	60	...	27	...	247	...	

INHERITANCE OF COAT-PIGMENTS AND COAT-PATTERNS

TABLE 11.—*Relation between distribution of pigment spots in 136 guinea-pigs of group D3, and in their parents.*

Offspring.	Parents.									Location of spots and total number of cases.
	4 spots.		3 spots.		2 spots.		1 spot.		0 spots.	
	Cases.	P. ct.	Cases.	P. ct.	Cases.	P. ct.	Cases.	P. ct.	Cases.	P. ct.	
2 spots.	117	89.3	4	80	...	...	...	...	...	...	⎫ Head; total cases, 136.
1 spot..	14	10.7	1	20	...	...	...	...	...	...	
0 spots.	...	...	...	...	...	...	...	...	...	...	
Total.	131	...	5	...	...	...	...	...	...	...	
2 spots.	...	...	2	...	...	...	1	...	48	36.6	⎫ Shoulder; total cases, 136.
1 spot..	...	...	1	...	...	...	...	...	28	21.4	
0 spots.	...	...	...	...	...	...	1	...	55	42.0	
Total.	...	...	3	...	...	...	2	...	131	...	
2 spots.	2	33.	...	...	38	52.8	...	...	19	32.8	⎫ Side; total cases, 136.
1 spot..	2	33.	...	...	14	19.4	...	...	13	22.4	
0 spots.	2	33.	...	...	20	27.8	...	...	26	44.8	
Total.	6	...	...	...	72	...	...	...	58	...	
2 spots.	113	83.1	...	...	...	...	...	...	...	...	⎫ Rump; total cases, 136.
1 spot..	14	10.3	...	...	...	...	...	...	...	...	
0 spots.	9	6.6	...	...	...	...	...	...	...	...	
Total.	136	...	...	...	...	...	...	...	...	...	

TABLE 12.—*Combination of the four parts of table 11, showing spot inheritance in group D3, 136 individuals (544 observations).*

Offspring.	Parents.										Total number of cases.
	4 spots.		3 spots.		2 spots.		1 spot.		0 spots.		
	Cases.	P. ct.	Cases.	P. ct.	Cases.	P. ct.	Cases.	P. ct.	Cases.	P. ct.	
2 spots...	232	85.0	6	75	38	52.8	1	...	67	35.5	⎫
1 spot....	30	11.0	2	25	14	19.4	1	...	41	21.7	⎬ 544
0 spots...	11	4.0	...	...	20	27.8	...	...	81	42.8	⎭
Total....	273	...	8	...	72	...	2	...	189	...	

TABLE 13.—*Combinations separately of the four parts of tables 5, 7, 9, and 11, showing inheritance of particular spots in series D as a whole, 461 individuals.*

Offspring	Parents.									Location of spots and total number of cases.
	4 spots.		3 spots.		2 spots.		1 spot.		0 spots.	
	Cases.	P. ct.	Cases.	P. ct.	Cases.	P. ct.	Cases.	P. ct.	Cases.	P. ct.	
2 spots.	336	87.0	15	60	35	70	...	...	...	...	Head; total cases, 461.
1 spot..	49	12.7	10	40	15	30	...	...	...	...	
0 spots.	1	0.3	..	...	...	...	...	...	...	...	
Total..	386	...	25	...	50	...	...	...	...	...	
2 spots.	4	80	10	47.6	13	31.7	9	45.0	114	30.5	Shoulder; total cases, 461.
1 spot..	1	20	7	33.3	11	26.8	5	25.0	87	23.2	
0 spots.	...	...	4	19.0	17	41.5	6	30.0	173	46.3	
Total..	5	...	21	...	41	...	20	...	374	...	
2 spots	15	55.6	7	77.8	78	48.4	8	50.0	85	34.3	Side; total cases, 461.
1 spot..	4	14.8	2	22.2	33	20.5	2	12.5	49	19.7	
0 spots.	8	29.6	..	...	50	41.0	6	37.5	114	46.0	
Total..	27	...	9	...	161	...	16	...	248	...	
2 spots.	306	80.7	...	...	41	50.0	...	...	...	...	Rump; total cases, 461.
1 spot..	39	10.3	...	..	20	24.4	...	...	...	...	
0 spots.	34	9.0	...	...	21	25.6	...	...	...	...	
Total..	379	...	...	...	82	...	...	...	...	...	

TABLE 14.—*Combination of tables 6, 8, 10, and 12, showing spot inheritance in series D in general, 461 individuals (1,844 observations).*

Offspring.	Parents.									Total number of cases.
	4 spots.		3 spots.		2 spots.		1 spot.		0 spots.	
	Cases.	P. ct.	Cases.	P. ct.	Cases.	P. ct.	Cases.	P. ct.	Cases.	P. ct.	
2 spots..	661	82.9	32	58.2	167	50.0	17	47.2	199	32.0	1844
1 spot....	93	11.7	19	34.5	79	23.7	8	22.2	136	21.9	
0 spots...	43	5.4	4	7.3	88	26.3	11	30.6	287	46.1	
Total....	797	...	55	...	334	...	36	...	622	...	

INHERITANCE OF COAT-PIGMENTS AND COAT-PATTERNS

TABLE 15.—*Relation between distribution of pigment spots in 207 guinea-pigs of group H_4, and in their parents.*

Offspring.	Parents.								Location of spots and total number of cases.
	4 spots.		3 spots.		2 spots.		1 spot.		0 spots.		
	Cases.	P. ct.	Cases.	P. ct.	Cases.	P. ct.	Cases.	P. ct.	Cases	P. ct.	
2 spots.	83	79.8	20	60.6	35	67.3	10	76.9	3	60.	Head; total cases, 207.
1 spot..	12	11.5	13	39.4	15	28.8	3	23.1	...	...	
0 spots.	9	8.7	...	...	2	3.8	...	...	2	40.	
Total.	104	...	33	...	52	...	13	...	5	...	
2 spots.	...	...	...	...	14	33.3	...	...	27	16.4	Shoulder; total cases, 207.
1 spot..	...	...	...	...	7	16.7	...	...	32	19.4	
0 spots.	...	...	...	...	21	50.0	...	...	106	64.2	
Total.	...	...	...	...	42	...	...	...	165	...	
2 spots.	...	...	...	...	20	40.0	1	9.1	25	17.1	Side; total cases, 207.
1 spot..	...	...	...	...	9	18.0	5	45.4	31	21.2	
0 spots.	...	...	...	...	21	42.0	5	45.4	90	61.7	
Total.	...	...	...	...	50	...	11	...	146	...	
2 spots.	...	...	...	...	44	52.4	10	34.5	26	27.7	Rump; total cases, 207.
1 spot..	...	...	...	...	17	20.2	8	27.6	14	14.9	
0 spots.	...	...	...	...	23	27.4	11	37.9	54	57.4	
Total.	...	...	...	...	84	...	29	...	94	...	

TABLE 16.—*Combination of the four parts of table 15, showing spot inheritance in group H_4, 207 individuals (828 observations).*

Offspring.	Parents.										Total number of cases.
	4 spots.		3 spots.		2 spots.		1 spot.		0 spots.		
	Cases.	P. ct.	Cases.	P. ct.	Cases.	P. ct.	Cases.	P. ct.	Cases.	P. ct.	
2 spots...	83	79.8	20	60.6	113	49.6	21	39.6	81	19.7	828
1 spot....	12	11.5	13	39.4	48	21.0	16	30.2	77	18.8	
0 spots...	9	8.7	...	...	67	29.4	16	30.2	252	61.5	
Total....	104	...	33	...	228	...	53	...	410	...	

IN RATS AND GUINEA-PIGS.

TABLE 17.—*Relation between distribution of pigment spots in 193 guinea-pigs of group H₁, and in their parents.*

Offspring.	Parents.									Location of spots and total number of cases.
	4 spots.		3 spots.		2 spots.		1 spot.		0 spots.	
	Cases.	P. ct.	Cases.	P. ct.	Cases.	P. ct.	Cases.	P. ct.	Cases.	P. ct.	
2 spots.	88	85.4	13	68.4	37	57.0	3	50.0	...	...	Head; total cases, 193.
1 spot..	11	10.7	5	26.3	21	32.3	1	16.7	...	...	
0 spots.	4	3.9	1	5.3	7	10.7	2	33.3	...	...	
Total.	103	...	19	...	65	...	6	...	...	...	
2 spots.	...	...	...	...	...	...	...	...	14	7.3	Shoulder; total cases, 193.
1 spot..	...	...	...	...	...	...	...	.	140	20.7	
0 spots.	...	...	...	...	..	...	...	...	39	72.0	
Total.	..	...	...	...	...	..	...	...	193	...	
2 spots.	...	...	...	..	...	...	...	..	29	15.0	Side; total cases, 193.
1 spot..	..	...	...	...	...	...	...	...	35	18.1	
0 spots.	..	...	...	...	...	...	...	...	129	66.8	
Total.	..	...	...	...	...	...	...	...	193	...	
2 spots.	...	...	...	...	...	...	...	...	54	28.0	Rump; total cases, 193.
1 spot..	...	...	...	...	...	...	...	...	31	16.0	
0 spots.	...	...	...	...	...	...	...	...	108	56.0	
Total.	...	...	...	...	...	...	...	...	193	...	

TABLE 18.—*Combination of the four parts of table 17, showing spot inheritance in group H₁, 193 individuals (772 observations).*

Offspring.	Parents.									Total number of cases.
	4 spots.		3 spots.		2 spots.		1 spot.		0 spots.	
	Cases.	P. ct.	Cases.	P. ct.	Cases.	P. ct.	Cases.	P. ct.	Cases.	P. ct.	
2 spots..	88	85.4	13	68.4	37	56.9	3	50	97	16.8	772
1 spot....	11	10.7	5	26.3	21	32.3	1	16.7	106	18.3	
0 spots...	4	3.9	1	5.3	7	10.8	2	33.3	376	64.9	
Total...	103	...	19	...	65	...	6	...	579	..	

44 INHERITANCE OF COAT-PIGMENTS AND COAT-PATTERNS

TABLE 19.—*Relation between distribution of pigment spots in 129 guinea-pigs of group H_2, and in their parents.*

Offspring.	Parents.									Location of spots and total number of cases.
	4 spots.		3 spots.		2 spots.		1 spot.		0 spots.	
	Cases.	P. ct.	Cases.	P. ct.	Cases.	P. ct.	Cases.	P. ct.	Cases.	P. ct.	
2 spots.	74	77.1	10	83.3	17	81.0	...	...	...	...	} Head; total cases, 129.
1 spot..	20	20.8	2	16.7	4	19.0	...	...	...	...	
0 spots.	2	2.1	...	...	...	...	...	...	...	...	
Total..	96	...	12	...	21	...	...	...	...	...	
2 spots.	..	...	...	...	...	...	...	...	4	3.1	} Shoulder; total cases, 129.
1 spot..	...	...	...	...	...	...	...	...	26	20.2	
0 spots.	...	...	...	...	...	...	...	...	99	76.7	
Total..	...	...	...	...	...	...	...	...	129	...	
2 spots.	...	...	...	...	...	...	...	...	11	8.5	} Side; total cases, 129.
1 spot..	...	...	...	..	...	...	...	...	30	23.3	
0 spots.	...	...	...	...	...	...	...	...	88	68.2	
Total..	...	...	...	...	...	...	...	..	129	...	
2 spots.	...	...	...	...	...	...	...	...	36	27.9	} Rump; total cases, 129.
1 spot..	...	...	...	...	...	...	...	...	17	13.2	
0 spots.	...	...	...	...	...	...	...	...	76	58.9	
Total..	...	...	...	...	...	...	...	...	129	...	

TABLE 20.—*Combination of the four parts of table 19, showing spot inheritance in group H_2, 129 individuals (516 observations).*

Offspring.	Parents.									Total number of cases.
	4 spots.		3 spots.		2 spots.		1 spot.		0 spots.	
	Cases.	P. ct.	Cases.	P. ct.	Cases.	P. ct.	Cases.	P. ct.	P. ct.	P. ct.	
2 spots...	74	77.1	10	83.3	17	81.0	...	...	51	13.2	} 516
1 spot....	20	20.8	2	16.7	4	19.0	...	...	73	18.8	
0 spots...	2	2.1	...	...	...	...	...	...	263	68.0	
Total....	96	...	12	...	21	...	...	...	387	...	

IN RATS AND GUINEA-PIGS. 45

TABLE 21.—*Relation between distribution of pigment spots in 58 guinea-pigs of group H3, and in their parents.*

Offspring.	Parents.								Location of spots and total number of cases.
	4 spots.		3 spots.		2 spots.		1 spot.		0 spots.		
	Cases.	P. ct.	Cases.	P. ct.	Cases.	P. ct.	Cases.	P. ct.	Cases.	P. ct.	
2 spots..	46	82.1	1	...	...	...	...	...	...	...	Head; total cases, 58.
1 spot ..	10	17.9	1	...	...	...	...	...	...	...	
0 spots..	...	...	...	...	...	...	...	...	...	...	
Total..	56	...	2	...	...	...	...	...	...	...	
2 spots..	...	...	...	...	...	...	...	...	2	3.4	Shoulder; total cases, 58.
1 spot ..	...	...	...	...	...	...	...	...	9	15.5	
0 spots..	...	...	...	...	...	...	...	...	47	81.1	
Total..	...	...	...	...	...	...	...	...	58	...	
2 spots..	...	...	...	...	...	...	...	...	4	6.9	Side; total cases, 58.
1 spot ..	...	...	...	...	...	...	...	...	14	24.1	
0 spots..	...	...	...	...	...	...	...	...	40	69.0	
Total..	...	...	...	...	...	...	...	...	58	...	
2 spots .	...	...	...	...	...	...	...	...	13	22.4	Rump; total cases, 58.
1 spot..	...	...	...	...	...	...	...	...	8	13.8	
0 spots .	...	...	...	...	...	...	...	...	37	63.8	
Total..	...	...	...	...	...	...	...	...	58	...	

TABLE 22.—*Combination of the four parts of table 21, showing spot inheritance in group H3, 58 individuals (232 observations).*

Offspring.	Parents.								Total number of cases.
	4 spots.		3 spots.		2 spots.		1 spot.		0 spots.		
	Cases.	P. ct.	Cases.	P. ct.	Cases.	P. ct.	Cases.	P. ct.	Cases.	P. ct.	
2 spots...	46	82.1	1	...	...	...	...	...	19	10.9	232
1 spot....	10	17.9	1	...	...	...	...	...	31	17.8	
0 spots...	...	...	...	...	...	...	...	...	124	71.3	
Total...	56	...	2	...	...	...	...	...	174	...	

46 INHERITANCE OF COAT-PIGMENTS AND COAT-PATTERNS

TABLE 23.—*Combinations separately of the four parts of tables 15, 17, 19, and 21, showing inheritance of particular spots in series H as a whole, 587 individuals.*

Offspring.	Parents.									Location of spots and total number of cases.
	4 spots.		3 spots.		2 spots.		1 spot.		0 spots.	
	Cases.	P. ct.	Cases.	P. ct.	Cases.	P. ct.	Cases.	P. ct.	Cases.	P. ct.	
2 spots.	291	81.1	44	66.7	89	64.5	13	68.4	3	60.	⎫
1 spot..	53	14.7	21	31.8	40	29.0	4	21.1	...	...	Head;
0 spots.	15	4.2	1	1.5	9	6.5	2	10.5	2	40.	⎬ total cases,
Total.	359	...	66	...	138	...	19	...	5	...	587. ⎭
2 spots.	...	...	...	...	14	33.3	..	...	47	18.6	⎫
1 spot..	...	...	...	...	7	16.7	...	...	107	19.7	Shoulder;
0 spots.	...	...	...	...	21	50.	...	...	391	71.7	⎬ total cases,
Total.	...	...	...	...	42	...	...	...	545	...	587. ⎭
2 spots.	...	...	...	...	20	40.	1	9.1	69	13.1	⎫
1 spot..	...	...	...	...	9	18.	5	45.4	110	20.9	Side;
0 spots.	..	...	...	...	21	42.	5	45.4	347	66.0	⎬ total cases,
Total.	...	...	...	...	50	...	11	...	526	...	587. ⎭
2 spots.	...	...	..	...	44	52.4	10	34.5	129	27.2	⎫
1 spot..	...	...	...	...	17	20.2	8	27.6	70	14.8	Rump;
0 spots.	...	...	...	...	23	27.4	11	37.9	275	58.0	⎬ total cases,
Total.	...	...	...	...	84	...	29	...	474	...	587. ⎭

TABLE 24.—*Combination of tables 16, 18, 20, and 22, showing spot inheritance in series H in general, 587 individuals (2348 observations).*

Offspring.	Parents.									Total number of cases.
	4 spots.		3 spots.		2 spots.		1 spot.		0 spots.	
	Cases.	P. ct.	Cases.	P. ct.	Cases.	P. ct.	Cases.	P. ct.	Cases.	P. ct.	
2 spots...	291	81.1	44	66.7	167	53.2	24	40.7	248	16.0	⎫
1 spot....	53	14.7	21	31.8	73	23.3	17	28.8	287	18.5	⎬ 2348
0 spots...	15	4.2	1	1.5	74	23.5	18	30.5	1015	65.5	
Total...	359	...	66	...	314	...	59	...	1550	...	⎭

TABLE 25.—*Frequencies with which spots absent in the parents (or in one parent, in group H½) are absent also in their offspring.*

	Group H½ (table 15).	Group H1 (table 17).	Group H2 (table 19).	Group H3 (table 21).	Average (table 23).
	Per ct.	Per ct.	Per ct.	Per ct.	Per ct.
Shoulder spots......	64.2	72.0	76.7	81.1	71.7
Side spots............	61.7	66.8	68.2	69.0	66.0
Rump spots.........	57.4	56.0	58.9	63.8	58.0

TABLE 26.—*Constants for the four groups of series D, based upon tables 6, 8, 10, and 12, respectively.*

	Group D½.	Group D1.	Group D2.	Group D3.
Average number of spots borne by a parent	3.78	4.40	4.62	4.64
Average number of spots borne by offspring	4.76	5.44	5.56	5.72
Per cent increase over parents	26	24	20	23
Standard deviation (σ) in number of spots, parents	2.876	3.812	3.744	3.694
Standard deviation (σ) in number of spots, offspring	3.503	3.615	3.663	3.662
Coefficient of variability, parents	76.78	85.32	80.90	79.44
Coefficient of variability, offspring	72.92	67.79	65.79	64.20
Coefficient of correlation between parents and offspring	0.4527	0.4580	0.4278	0.4343

TABLE 27.—*Constants for the four groups of individuals of series H, based upon tables 16, 18, 20, and 22, respectively.*

	Group H½.	Group H1.	Group H2.	Group H3.
Average number of spots borne by a parent	2.48	1.56	1.80	2.00
Average number of spots borne by offspring	3.88	3.20	3.08	3.00
Per cent increase over parents	56	105	71	50
Standard deviation (σ) in number of spots, parents	2.870	2.922	3.198	3.434
Standard deviation (σ) in number of spots, offspring	3.5768	3.6076	3.5960	3.6200
Coefficient of variability, parents	115.72	186.57	177.66	171.70
Coefficient of variability, offspring	92.18	112.75	115.00	120.60
Coefficient of correlation between parents and offspring	0.4565	0.6076	0.6109	0.6796

TABLE 28.—*Character of the young of the original nose-spot male, 1989a (NS_0).*

	Individuals with nose spots.	Total other spots.	Average number other spots.	Individuals without nose spots.	Total other spots.	Average number other spots.	Per cent of young with nose spots.
By mothers without nose spots	4	6	1.5	14	40	2.8	22.2
By NS_0 mothers	22	58	2.6	19	58	3.0	53.6
By $NS_{\frac{1}{2}}$ mothers	16	39	2.4	10	29	2.9	61.5
By NS_1 mothers	3	5	1.6	2	1	0.5	60.0
Total	45	108	2.4	45	128	2.8	50.0

48 INHERITANCE OF COAT-PIGMENTS AND COAT-PATTERNS

TABLE 29—*Character of the young of nose-spot males, chiefly descendants of 1989ª (table 28).*

	Individuals with nose spots.	Total other spots.	Average number other spots.	Individuals without nose spots.	Total other spots.	Average number other spots.	Per cent of young with nose spots.
♂5595, NS₀, by NS₀ mothers	1	5	5.	6	27	4.5	14.3
" " , " NS₁ "	4	10	2.5	5	15	3.	44.4
♂5678, NS₁, " NS₀ "	0	0	0.	2	5	2.5	0.
" " , " NS₁ "	2	6	3.	3	8	2.7	40.
♂5151, NS₂, " NS₀ "	15	30	2.	22	66	3.	40.5
" " , " NS₁ "	9	32	3.5	13	52	4.	40.9
♂5652, NS₂, " NS₀ "	7	16	2.3	5	13	2.6	58.3
" " , " NS₂ "	2	3	1.5	2	9	4.5	50.
♂5388, NS₂½, " NS₀ "	0	0	0.	4	14	3.5	0.
" " , " NS₁ "	5	6	1.2	6	11	1.8	45.4
♂5669, NS₂½, " NS₀ "	1	3	3.	4	10	2.5	20.
" " , " NS₁ "	2	5	2.5	1	1	1.	67.
" " , " NS₂ "	1	5	5.	1	4	4.	50.
Total	49	121	2.5	74	235	3.2	39.8

TABLE 30.—*Further analysis of the data contained in table 29.*

	Individuals with nose spots.	Total other spots.	Average number other spots.	Individuals without nose spots.	Total other spots.	Average number other spots.	Per cent of young with nose spots.
NS₀ males by NS₀ females	1	5	5.0	8	32	4.0	11
NS₀ males by NS₁ females	6	16	2.7	8	23	2.9	43
NS₁ or 1½ males by NS₀ females.	23	49	2.1	35	103	2.9	39.6
NS₁ or 1½ males by NS₁ females.	16	43	2.7	20	64	3.2	44
NS₁ or 1½ males by NS₂ females.	3	8	2.7	3	13	4.3	50
Totals for NS₀ males	7	21	3.0	16	55	3.4	30.4
Totals for NS₁ or 1½ males	42	100	2.2	58	180	3.1	42
Totals for NS₀ females	24	54	2.1	43	135	3.1	35.8
Totals for NS₁ females	22	59	2.7	28	87	3.1	44
Totals for NS₂ females	3	8	2.7	3	13	4.3	50
Grand totals	49	121	2.5	74	235	3.2	39.8

IN RATS AND GUINEA-PIGS.   49

TABLE 31.—*Character as regards nose spots of the young of animals of Series H (head spot series). None of the parents bore nose spots.*

Father.	Spots of father.	Spots of mother (average).	Spots of parents (average).	Individuals with nose spots.	Total other spots.	Average number other spots.	Individuals without nose-spots.	Total other spots.	Average number other spots.	Per cent of young with nose spots.
1716*	1	2.0	1.5	4	16	4.0	29	100	3.4	12.1
2537	0	2.5	1.25	10	19	1.9	63	197	3.1	38.7
2625	1	1.9	1.5	14	32	2.3	47	132	2.8	23
4388	0	3.2	1.6	4	7	1.7	34	94	2.8	10.5
4399	1	2.7	1.8	7	14	2	21	48	2.3	25
5420	1	1.7	1.3	...	...	...	6	18	3.0	0
Total.	0.56	2.3	1.4	39	88	2.3	200	589	2.9	16.3

*By an NS_0 female, this same male (1716) had ten young, three of which bore nose spots. The average number of spots other than nose spots borne by the NS young was 2.6; by the remaining young, 3.7.

Data for curves (figs. 2 to 5).

Classes...	4.5	14.5	24.5	34.5	44.5	54.5	64.5	74.5	Average.
A.........	75	64	35	9	...	...	...	...	13.3
B.........	29	35	34	15	6	3	4	...	21.2
C.........	8	8	7	9	2	1	1	...	23.5
D.........	45	53	29	9	15	17	11	1	23.9
E.........	24	48	29	16	10	5	...	...	21.1
F.........	32	42	12	3	...	...	...	...	12.9
Classes...	2	7	12	17	22	27	32	...	...
G.........	28	17	17	12	6	2	1	...	9.65
H.........	31	20	7	2	1	...	...	...	5.6
I.........	24	6	4	...	...	...	...	...	4.
Classes...	4.5	14.5	24.5	34.5	44.5	54.5	64.5	74.5	...
J.........	34	44	21	10	1	1	...	...	15.76
K.........	16	18	16	9	9	1	1	...	22.2
L.........	4	9	14	10	9	11	4	3	36.2
M.........	...	...	1	2	4	6	5	1	52.4

www.ingramcontent.com/pod-product-compliance
Lightning Source LLC
Chambersburg PA
CBHW031322150426
43191CB00005B/303